LAND HEALER

LAND HEALER

How Farming Can Save Britain's Countryside

Jake Fiennes
with Tim Ecott

3

Witness Books, an imprint of Ebury Publishing,
20 Vauxhall Bridge Road,
London SW1V 2SA

Witness Books is part of the Penguin Random House group of companies
whose addresses can be found at global.penguinrandomhouse.com

First published by Witness Books in 2022
Paperback edition published in 2023

www.penguin.co.uk

A CIP catalogue record for this book is available from the British Library

ISBN 9781785947315

Printed and bound in Great Britain by Clays Ltd, Elcograf S.p.A

The authorised representative in the EEA is Penguin Random House Ireland,
Morrison Chambers, 32 Nassau Street, Dublin D02 YH68

Contents

Jake Fiennes is Head of Conservation for the 25,000-acre Holkham Estate in North Norfolk. With a thirty-year career in land management, his advice is sought after by key players in conservation and agriculture – including King Charles III of England, the Department for Environment, Food and Rural Affairs (DEFRA), the NFU, the National Trust, the RSPB and Natural England, among others. Previously, Fiennes worked at the Raveningham and Knepp Castle estates, building up his extensive experience and understanding of how farming impacts our natural environment. He demonstrates how, with the right approach, we can rebuild biodiversity while still producing food.

For those with mud on their boots
and calluses on their hands

I

Hedge Porn

June
Great Farm, North Norfolk

Yellow. A flash. Quick short wingbeats, one, two, three . . . four . . . five. Glide. Dip. Wingbeat . . . one, two, three, four . . . and rise again. Another power glide. And into the hedge. It's gone. A male yellowhammer, no bigger than a sparrow but instantly recognisable for its lemon crown. The hedge is high, twelve feet or more where the land slopes downwards and just a touch to the south-west on this patch of chalky soil.

I'm sitting in my 4x4 where the ground slopes away gently at the corner of Tinkers Hole and the Twenty Acres, two of the fields at the heart of Great Farm. From this spot I can see only the fields, the hedgerows and the wide Norfolk sky. The air is pleasantly warm and the field margins are filled with wildflowers. Tall dark beacons of viper's-bugloss with bright blue spiky flowers stand proud of the undergrowth, and here and there the pale spires of wild mignonette, the delicate yellow plant used in ancient times to make dye, wave gently in the light breeze. Below them, in the tangle of foliage just a few inches high, the shorter, stumpier, common mallow makes a show with its eye-catching fuchsia-pink petals. There are pure white ox-eye daisies by the thousands, and bees, hoverflies and dragonflies create a low rumbling soundscape. On patchy ground near the hedgeline, I spot night-flowering catchfly, sometimes called the 'clammy cockle', a rare plant that flowers during the dark and releases its scent to attract moths. It's only in flower for a month or so, and not very spectacular, but it has five tiny double-lobed petals that look like bunny ears.

Here, there is a sense of abundance. The flowering plants have made a living carpet of colour, attracting hosts of insects for the birds to feast upon. This is the time of greatest variety, and when there's maximum food in the fields. That means it's also when most of the birds are nesting, laying and rearing their chicks. The track where I'm sitting is new, and it too has its role to play in the way this farm welcomes nature. In the tramlines made by the wheels of farm vehicles, a female grey partridge scurries ahead in a dead straight line on the scrubby ground. After a second or two I see five tiny chicks emerge from the long grass at the base of the hedge and follow her in single file. They like to forage for insects on paths where the undergrowth isn't too thick, ducking into the crops or the hedgeline when she calls an alarm.

On the far side of Tinkers Hole I can see brown hares using the 13-hectare field as a giant boxing ring. Five, six – no, wait – seven animals are running back and forth on the open ground. They make short-lived pairs, circling and breaking apart as individual males seek to challenge one another and disrupt other courting couples. They dance and weave across the lower half of the field, ignoring me so long as I don't open the truck door and step out. High above it all, there is a female marsh harrier gliding over the open ground but the hares carry on. It's mating time and very little will interrupt them. And the harrier, in spite of her impressive four-foot wingspan, would be more likely to take a leveret than attack a healthy adult hare anyway.

Right now, my hedge is a safe place for the little yellowhammer. I keep watching and soon see more of these tiny birds making their way in and out of the foliage. In the space of a couple of minutes, I count seven or eight males coming and going. One bird perches on an arching bough, a bramble frond that bounces gently like a rocking horse under its weight. Then comes the song. The distinctive call of the yellowhammer is famous, the rapid half-dozen cheeps followed by the last, drawn-out note that countrymen say sounds like '*a little bit of bread, and no cheese* . . .' There are linnets, too, flying in groups with a call that is higher pitched – *chi-chi-chi-chi-chewchipchipchew* – and even smaller than the yellowhammers, with a similar undulating flight but easily singled out by their pale raspberry-tinged breast feathers, as if they have been supping from a jam jar.

This is the time of year when every farm in the country should be looking its best. The hedge has more growing to do, but this year it's putting its energy into producing berries. It will get bigger and bolder with each successive growing season, and then it will reach a natural limit. Meanwhile, there's so much happening on the ground now: pink campion and musk mallow and shepherd's purse. There is a stand of tall maiden's tears – also known as bladder campion – the dainty fine petals emerging above its sack-like calyx. Crossbills have been spotted here, their overlapping beaks like scissors that have gone wrong, but designed perfectly for picking the seeds out of pine cones. The farm is buzzing and alive with a mass of colour and texture.

Eighteen months ago, when I arrived here, there were no yellowhammers or linnets nesting on the edge of this field. There were scarcely any breeding pairs of birds in evidence. This land had been farmed hard by a tenant for many years. It was in a rigorous cycle of root-crop rotations, and the soil had been ploughed right up to the hedgeline. The hedge itself was kept at six foot high, and brutally trimmed so that all the hedgerows on the farm were uniformly 'neat'. But, since then, the hedge between Tinkers and the Twenty Acres has been left untrimmed. It's dome shaped, not cut like a 'flat top' hairstyle, and now contains many different types of plants. After two seasons without being harshly cut with a mechanical flail, there are now fruits and seeds aplenty for the birds that nest in it. It's got blackthorn and hawthorn which flower at different times, and brambles which will produce both flowers and fruit in a month or two. Aside from biodiversity for plants and insects, hedges are a useful shelter for livestock, and the cattle often browse on hazel, hawthorn and blackthorn, obtaining minerals from these plants which may be lacking in grass.

The simple logic is that there's no good reason to cut this hedge. It's not beside a road, so it isn't obscuring any vehicle sightlines, nor is it taking up space that could be used for crops. Its size is important too, as only after a certain height does it produce enough cover and enough shelter from the rain to make it a safe home for birds to rear their chicks successfully. Songbirds like a high hedge – where they can sit and sing when they are marking out their territory and calling for a mate. I bring as many farmers here to this spot as I can. If you love hedges, then this one is what I jokingly call a prime example of 'hedge porn'. It does what a hedge should do. It is the very image of what a hedge should look like.

Before we started managing the land at Great Farm in a more nature-friendly way, it was costing over £2,500 a year to cut these hedges. That might not sound like a lot of money, but it's a significant saving when, according to Defra (the Department for Environment, Food and Rural Affairs) the average yearly income of all farm types in the UK in 2020 was £46,000. Within that, some £27,800 would have come from the Basic Payment Scheme, a subsidy system designed to give guaranteed income to every farm in the EU, and which replaced earlier subsidies paid to those who produced particular crops. It's been a crucial part of the British agricultural system for many years. Perhaps unsurprisingly, a recent survey by the National Farmers' Union (NFU) found that almost 90 per cent of farmers were in favour of hedgerow restoration as a simple carbon-cutting measure. It's an easy win, but many landowners and farmers are still fixated on making things look 'tidy', whatever the environmental cost.

There are around half a million miles of hedgerow in England; that's almost twice as much as the length of all the tarred roads, and by some calculations makes them our largest single nature reserve. Of the roughly 1,100 'priority species' in the UK, more than 10 per cent of them are significantly associated with hedgerows. But that's still less than half the amount of hedgerow that existed a century ago, and about a third of what existed in the 1950s, before everything in farming became supercharged by the need to make the industry efficient. Yet, the Environment Secretary recently called them 'the single most important ecological building blocks in our farmed landscape'. Others have called them the 'stitching in the countryside quilt'. In terms of the economy, one recent estimate claims that for every £1 spent on hedgerows, they will return £3.92 in ecosystem services and the economic activities associated with them. These include the enhancement of biodiversity, carbon sequestration and the provision of jobs and biofuels. For too long, too many farmers have treated hedges as if they are simple boundary markers, a fence that needs trimming to maintain an orderly shape. They have not treated them as living organisms that create habitat, refuge and forage, bringing wide benefits to the environment.

The government defines a hedgerow as a boundary line of trees or shrubs extending more than twenty metres, but less than five metres wide, and having no gaps more than twenty metres long. It's important that these things are defined, so that they can be measured and assessed. Recent estimates say those hedges around the whole country will store around 9 million tonnes of carbon. The Climate Change Committee and the countryside charity CPRE (formerly the Council to Protect Rural England) are both urging the government to increase our network of hedgerows by 40 per cent by the middle of the century in order to help achieve the target of net zero carbon emissions. It's already calculated that about 5 per cent of the carbon sequestration on English farms is carried out by existing hedgerows. Meanwhile, the independent Organic Research Centre calculates that one hectare (2.47 acres) of this habitat (between 3.5 and 6 metres wide) will sequester more than 130 tonnes of carbon per year. Food production and distribution account for about a fifth of the UK's greenhouse gas emissions, and 40 per cent of those come from agriculture. As if that evidence isn't compelling enough, the UK Centre for Ecology and Hydrology calculates that without hedgerows, pollinator numbers on farmland might be reduced by one-fifth. The diversity of plants within a hedge are its key attraction, and when they sit alongside intensively managed land they become even more important.

The loss of our historic patchwork of hedges (you might think of them as the veins of the countryside) is the legacy of the drive for more productive farming over the past seventy years. And it's not just the hedges and the wildflower meadows; we've also lost half of our ancient woodlands and 90 per cent of lowland ponds. Why? Because for most of the twentieth century, more yield per hectare became the mantra. In the Second World War, the amount of land used for arable production in the UK increased by 50 per cent to almost 8 million hectares (20 million acres). Up until then Britain was only producing about 30 per cent of its own food, but by the end of the war the nation was growing 75 per cent of what it needed.

Since then the trend has been downward, and it's now only about 6 million hectares (15 million acres). However, from the end of the war up until the millennium, yields per acre for things like wheat, oats and barley tripled. Globally, the story is very similar. According to the 2021 Dimbleby report, a review conducted for the government into the national food strategy, in this new Green Revolution we are producing 1.7 times as much food as in 1960, on about a third of the land.

That might sound like a good thing, but to make fields work harder and produce those higher yields it has been increasingly necessary to add chemicals both to the land and to the growing crops. Anything that competed with the crops – animals or other plants and insects – was classified as pest, vermin, weed or blight. The irony of that way of thinking is that hedgerows, all too often ripped out in the name of 'efficiency', were a key reservoir of predators that kept pests under control in the crop fields. In winter, they sheltered livestock from the wind, and in summer they gave them shade. They provide a long list of benefits, or to use the current terminology – they increase the 'delivery of ecosystem services'. Getting rid of so many of them was a disaster.

Perhaps more seriously damaging than the chemical and mechanical assault on the land, in this post-war journey to maximum production, has been the shift in the way we think about the land and what lives on it. When I was talking to a farmer recently about how we could incorporate a wildflower meadow in a strip around some of his fields to increase biodiversity, he was sceptical. I pointed out that the insects the flowers attracted would support great bird numbers, and he would be rewarded with government payments for a stewardship scheme aimed at helping nature. The idea of growing something which would not directly be part of what he could sell was hard for him to swallow. 'But that's not what I'm used to doing,' he said. 'I have spent my whole career killing the things I don't need!'

On Great Farm, I've made some small changes that allow the land to remain in food production, but which also allow nature to thrive. Not just survive, but prosper. With a bit of forethought it's not very difficult to achieve. And I believe that pretty much every farm in Britain can do the same, without damaging the amount of food we produce. This doesn't mean turning the land into some kind of artificial wildlife park. This book is about how to bring about small changes that could save the countryside for farming and for nature. I've spent over thirty years working on the land and I'm optimistic about the way nature can recover, despite all the terrible things we've done in the name of efficiency and modernisation.

Anyone who's interested in restoring the natural environment and redressing the balance of nature so that wildflowers, birds, amphibians, fish, insects, other invertebrates and indeed the soil itself can thrive, needs to understand some basic facts about the way the countryside works. They should also understand that over 70 per cent of the UK's land area is used for agriculture. That's almost 19 million hectares (47 million acres), with just over a third of it (i.e. 25 per cent of the total country) given over to arable farms, with much of the remainder being grassland or pasture for livestock. Around one-third of our total agriculture area is used for crops, with a further 5 per cent for woodland. Defra says that cereals – mainly wheat and barley – account for half of the croppable area. The farmland is home to about 9.6 million cattle (beef and dairy), 32.7 million sheep and lambs, just over 5 million pigs and about 180 million poultry (of which chickens are almost 120 million). That agricultural land is divided up into around 180,000 farms, employing just shy of half a million people.

In Britain, the number of people who work the land has been steadily decreasing, partly due to increased mechanisation and partly due to a dearth of young people wanting to get into farming. In 2020 there were officially 107,000 self-employed farmers, a sizeable reduction from 138,000 the previous year; as recently as 2014 there were as many as 167,000 farmers in the country. The trend is ever downward. Those farming businesses still going are now mostly being run by farmers with an average age of just below sixty – and that's already come down significantly in the last few years. Barriers that stop younger people taking over farms – including the price of rural housing and low farm wages – are another factor in why some of the less enlightened practices have persisted so long.

It should be pretty obvious that if you want to improve our natural environment, you need to talk to our farmers. And, when you think about some of those big numbers, consider that there are just over two hundred national nature reserves in England, covering less than 100,000 hectares (slightly more than a quarter of a million acres). That's only about 0.7 per cent of the country. Much as I hate to say it, we're not going to do very much for improving national biodiversity and saving endangered species by putting our efforts into those small areas. It's probably not a good idea to think of restricting our wildlife, whether it be plants, animals or insects, to these patches. It would result in a few isolated islands of 'life', which would not be much better than creating zoos. It's pretty clear that the natural world is not something we 'add on' to our daily lives when we feel like it. We actually depend on it for our survival.

For many people, especially the older generation, an idyllic memory of what the English countryside is like persists. It's embalmed in literature and films, triggered by hearing the theme tune of *The Archers* and by half-remembered children's books. Whether it's *Tales from Beatrix Potter*, Enid Blyton's *Famous Five* or Roald Dahl's *Danny, the Champion of the World*, these stories are cultural pillars that have become deeply engrained in our vision when we think about the countryside. Adult literature is equally powerful, even when it conjures darker themes about the harshness of an agricultural world in Thomas Hardy, its humorousness as described in Stella Gibbons's *Cold Comfort Farm*, or the bucolic innocence of village life as it confronts modernisation in Laurie Lee's *Cider with Rosie*. Other seminal works have imbued the essential qualities of individual species in imagination, from the cheekiness of otters in Gavin Maxwell's *Ring of Bright Water* to the free-spiritedness of hunting birds in Barry Hines's *A Kestrel for a Knave* or their rapacious efficiency as detailed by J.A. Baker's country diary of *The Peregrine*. These books, often capturing a wider, multi-generational audience by being turned into films, all contribute to a slow drip feed that has built a folk memory of 'the country' that we hold dear.

What people don't always understand is that when our landscapes were rich in nature and embodied some ideal of 'Englishness', it was *because* of farming, not in spite of it. In his seminal *History of the Countryside*, Oliver Rackham points out that there is a dangerous fallacy, known as the 'Kaleidoscope Myth', which asserts that we tend to love the things we are familiar with and hold features in the landscape dear that we have 'always' seen. However, he believes we should not be afraid of change, because it has always been present. In most cases those familiar features are the result of how the land has been farmed. Rackham observes that 'except for town expansion, almost every hedge, wood, heath, fen etc. on the Ordnance Survey maps of 1870 is still there on the air photographs of 1940.' It's worth bearing that in mind when we try to comprehend how big the changes in our countryside environment have been since the end of the war. At that time, when my parents were children, 'Much of England,' Rackham observes, 'would have been instantly recognizable by Sir Thomas More.'

However, what's changed since 1945 isn't necessarily the static aerial view of our patchwork of fields and rivers, it's the balance of living things portrayed in the much-loved books and films I mention. That vibrancy and the element of surprise in spotting birds or animals or flowers as we walk about the countryside has largely disappeared, although there is arguably still a community of people who think of themselves as country folk and who hold dear a set of values and lifestyle that differs quite radically from what they regard as urban or metropolitan. However, the way the country community works and earns its living off the land has also changed in very significant ways. Somehow, slowly and stealthily, while we were in pursuit of productivity and profit, we jumped on a treadmill of ever-increasing yield and input. Hay meadows, the quintessential symbol of what the English countryside has lost, were very much part of a farmed system – yet they provided a rich and varied diet for our livestock whilst also benefiting nature. Since the

1930s, however, over 97 per cent of those meadows have disappeared – equivalent to an area one and a half times the size of Wales. Harebells, ragged robin and field scabious, bird's foot trefoil, wild red clover and spider orchids have largely gone with them, flowers that were easily recognisable for a country child back then.

Without the cooperation and positive engagement of farmers, any attempt at redressing some of the environmental damage we have done to the countryside will be futile. To those who want to protect the things that make the English landscape special, my message is simple: farmers are not the enemy. I think we also need to remember that the countryside we think we want to 'preserve' is not necessarily natural – it's been managed and shaped and bent to our needs for thousands of years. In other words, we have been changing and affecting what goes on in the countryside for a very long time. But in recent decades we have intervened in a more targeted and radical way, especially when it comes to allowing wild species – or those classified as 'non-commercial' – to survive within our farmed environment. There is, at times, an irrational fear of 'weeds' and 'invasive' plants or pests, which can often be effectively managed or made less of a 'problem' by changing the time of year at which we do certain things, rather than always reaching for the arsenal of chemicals that the agrichemical industry provides. That habit needs to change.

The evidence for species decline and the degradation of our landscape is grim. By some measures, England has the most depleted natural environment in Europe. That description comes from the UK's largest partnership for nature, the National Biodiversity Network, which uses data collected over several decades by membership organisations including the RSPB, the Natural History Museum, Natural England and the Environment Agency, among many others. The Natural History Museum's own *Biodiversity Intactness Index* (2021) estimates the percentage of the original number of a species that remain and their abundance in any given area, despite human impacts. The index puts the UK at just 53.2 per cent, with a predicted rise of 2 percentage points by 2050. The museum reiterates that the decline in animal and plant species is not simply an environmental issue, but is linked inextricably to development, as well as to economics, ethics and morals. They underline that the biggest loss of biodiversity has been caused by changes in land usage, especially unsustainable agriculture. A relatively small percentage of land loss is also due to increasing residential and commercial development.

Under these scenarios, and with this mountain of evidence, it's clear that making space for nature within a farming context is one of the most pressing issues with which we are faced. But, we don't need to throw up our hands in horror and resign ourselves to disaster. The good news is that nature has remarkable powers of recovery. And it can happen quickly if you give it a small helping hand.

Nonetheless, there's a lot of bad news out there. It's tiring and demoralising and sometimes seems so big a problem that we, as individuals, lose hope. Many environmentalists, farmers, and even governments seem to think that protecting our natural world is such a big challenge that it will eventually only be resolved by someone else, or by the arrival of some new kind of technology. They seem to be pinning their hopes on an as yet undiscovered magic bullet. Perhaps something like that will come along, but there's no sign of it, and it's a bit like someone who needs money to pay the rent pinning all their hopes on winning the Premium Bonds. Instead of waiting for this as yet unknown magical solution, it seems much more sensible to get everyone to do their bit to make things better right now. I can show, and have already demonstrated within a relatively short timescale, that on Great Farm, and on thousands of others like it around the country, there is much to be hopeful about. Fixing a lot of what has gone wrong in our natural surroundings isn't overly complicated. Don't get me wrong – it's a big job; it involves preserving what we love in the natural world as well as supporting a healthy agricultural sector, which is an essential part of the rural economy. Making things better is an urgent priority if we are going to maintain healthy wildlife habitats. And, I believe there's also a lot more healthy 'nature' out there, surviving and ready to flourish, than many people think. It doesn't matter if we are birdwatchers out walking at weekends or working farmers in our insulated tractor cabs navigating our way around the fields with the aid of the latest GPS technology, we all need to open our eyes and look at what's around us. We need to really notice what nature is telling us, and give it that assistance, especially when doing that won't stop us working the land and producing the food our constantly increasing population needs. To put it crudely, I firmly believe that the natural world is *not* totally fucked. We can fix it.

2
Simple Truths

Great Farm sits at the western edge of the Holkham Estate, the heart of the ancient properties acquired by the Earls of Leicester. For centuries, their land has covered a large area of north Norfolk, a healthy chunk of East Anglia where the rump of England bulges into the North Sea. The grand family seat of Holkham Hall is surrounded by a vast park encircled by a flint and brick wall nine miles long. The parkland is carefully laid out and planted with more than a million trees. It has all the accoutrements of the classic English country estate, including features designed by William Kent, Capability Brown and, more significantly, Humphrey Repton: long tree-lined avenues, vast stands of weathered oaks, walled vegetable gardens, roaming herds of fallow deer, a classical temple, a quirkily shaped thatched ice-house and an impressive boating lake at the centre of it all. The working farmland and buildings stretch from Stiffkey to the east some ten miles westwards all the way to Burnham Norton. From north to south the estate goes almost twelve miles inland from the high-water mark at Holkham beach to somewhere around the village of

South Creake. In all, the modern estate covers around 25,000 acres (just over 10,000 hectares), some of it farmed by the estate itself, but mostly given into the charge of tenant farmers.

Holkham is one of England's greatest private houses and incorporates an important national nature reserve covering over 3,700 hectares (just over 9,000 acres), a panoramic stretch of sandy beach (where the Household Cavalry come each year to give their horses a holiday) and miles of brooding, boggy saltmarshes, as well as thousands of acres of commercial farmland, most of it arable. In late 2018, Tom Coke, the current Lord Leicester, made me Director of Conservation for the estate. His family have lived here since the early seventeenth century, descendants of the prominent Elizabethan lawyer Sir Edward Coke (1552–1634), who as Lord Chief Justice prosecuted both the men behind the Gunpowder Plot and Sir Walter Raleigh. He amassed considerable wealth and land, but later ran foul of King Charles I by introducing the Petition of Right, which challenged the monarch's attempts to impose a more autocratic rule.

Construction on the imposing family seat, Holkham Hall, began in 1734 (a century after Sir Edward's death) and was completed thirty years later by Thomas Coke, 1st Earl of Leicester. Always known simply as 'Coke of Norfolk', the first earl was famous for his hospitality and his energetic interest in country matters. He is credited with numerous agricultural innovations, many of which were well ahead of their time. Presciently, many of the improvements he made then would chime well with what we now call sustainable farming. The big house is an architecturally notable Palladian villa with a façade more than a hundred yards wide and set in over three thousand acres of formal parkland. Inside there are furnishings of velvet and rich brocades, gilt mouldings, grand paintings and classical sculptures. The marbled entrance hall is modelled on a Roman temple and has a ceiling fifty feet high. It's grand in every sense of the word.

Great Farm is not grand. And, unlike much of the estate, it's only been owned by Holkham since 1922, when it was part of a large parcel of land acquired from the last Earl of Orford. It was one of the final major land acquisitions by the estate after the First World War. This was the era when many of the historic English landholdings were famously broken up and sold off as the aristocratic owners found it impossible to afford their upkeep, and there was increasingly a shortage of workers to maintain them. At the top of a narrow staircase, somewhere above the great kitchen, the Hall archives hold the deeds and records of these land purchases. Going back as far as Anglo-Saxon times, almost every field in England has a name, and the archives contain a leather-bound ledger containing the original details of Great Farm. Inside, a faded, cracked map has been worn into holes by folding and refolding, and by enquiring fingers jabbing at the page to point out one feature or another.

The Orford map shows Great Farm, a patchwork of fields laid out south of Burnham Deepdale and west of

Burnham Norton. Each one is neatly numbered and cross-referenced in copperplate script with their exact acreage recorded in the ledger. Tinkers Hole is there, marked as land parcel 'Number 14', and listed then as being just over 37 acres in area. The adjoining plots are named as Brick Kiln Piece (34 acres), High Field – also known as the Twenty-One Acres – and the Fifteen Acres. There is also a Chalk Pit (1.5 acres), which became Bottle Pit in the twentieth century when it was a place for dumping glass in the days before recycling. At the far western edge of the farm there is the irregular shape of Bullock Shed Close, which some time over the next half-century became known as Brick O'Longs. Many of the local Norfolk names are utilitarian, and only occasionally more descriptive, like Hudson's Marsh, or more intriguingly antique, like Handcups Piece, Starve's Crow or Knave's Acre. At Holkham, near Burnham Market there is a field called Midnight, and the theory goes that it has some connection to how long it used to take to walk back from there to the farm. The very oldest names are often impossible to divine, but the sea wall at Overy has a section called Whincover, derived from the old Norse word *hvin* meaning 'gorse', and indeed that section is still lined with yellow gorse bushes.

Much of the landscape predates the formal parkland around Holkham Hall and was probably used by farmers and pastoralists for thousands of years before the estate was created. Clearly, Great Farm was never parkland laid out with ornamental lakes and decorated with monuments. It was a typical North Norfolk farm, producing root crops like sugar beet, and potatoes, and grains like wheat and barley.

Great Farm was one of the first places I visited when I arrived at Holkham in late 2018. The weather was cold and there was very little life around. The landscape itself was in a fairly sorry state. An environmental survey of the farm was anything but encouraging. It was described as having 'poor to average biodiversity'. Its soil was depleted and there was an absence of cover crops. The woodland areas were overcrowded, dominated by spruce and badly in need of thinning. There was a lack of margins suitable for invertebrates. The 106 hectares (260 acres) had just been vacated by a commercial tenant who had worked the ten fields within its boundaries intensively for some years. The soil on the farm is a medium loam – a mixture of sand, silt and clay, but the site is exposed to northerly winds. And here, so close to the North Sea, those winds can be strong and bitter.

The records show that Great Farm had been in a fairly relentless cycle of commercial production. From 2010, the High Field was planted successively with potatoes one winter, then winter wheat, then sugar beet, then winter barley, then parsnips, then maize, then sugar beet and back to potatoes again, all of which suck nutrients from the soil and don't replenish it. All of the fields had similar rotations, the land worked hard with what farmers now call 'inputs' – fertiliser, herbicides, pesticides – often with multiple applications of each at different stages of the crop cycle. Sadly, it's not an unusual or particularly unique story. But using the land in that way doesn't allow time or space for nature. The crops were planted as close to the field edges as the farm machinery could go, and the hedges were cut short and stumpy for 'neatness'. The last thing I saw being lifted from the soil was the sugar beet, which was riddled with slugs and leatherjackets, pests that often flourish when the crops are planted in too frequent rotations.

The kinds of standard farming practices I've described can deliver profitable yields from the crop cycle. But in many cases that profit is only made possible because of the subsidies that have been paid to farmers over the last several decades by the government, mainly under the terms of the European Union's Common Agricultural Policy. When the UK left the European Union, those subsidies amounted to around £3.5 billion per annum for our farmers. Sadly, it means that many landowners have been effectively 'farming subsidies' just as much as they have been farming their fields. The net value of the crops after the running costs of the farm, labour and equipment have been taken into account mean that the profit was just about equal to the subsidy.

The original idea behind the CAP subsidies was laudable enough, aiming to improve agricultural productivity by promoting technical progress and ensuring that farmers were paid a fair amount for the vital job they do. It was also aimed at ensuring that markets for food products remained stable and less subject to fluctuations in weather. It was also aimed at making prices stable for us, the consumers. In the 1970s, however, the subsidy system was responsible for a series of gluts. Originally intended to rectify shortages in certain products, or to smooth out annual variations in levels of production, the EU subsidies encouraged farmers to produce certain commodities because they knew they could get a guaranteed price. This, notoriously, resulted in the so-called EU 'butter and beef mountains' and 'wine and milk lakes'.

One of the solutions to reducing that obvious waste of natural resources was to introduce the practice of 'set-aside' (in place from 1988 to 2008), where farmers were paid to remove a percentage of their fields from production. Although those areas may have shown improved biodiversity in some cases, they were just an unintended consequence of the set-aside scheme: the actual purpose of leaving the land unplanted was merely to reduce the production surplus. It wasn't a conservation measure *per se*.

In 1973, when the UK joined what was then known as the Common Market, agriculture was about 80 per cent of the bloc's budget, although that went down considerably over the years and was about 40 per cent when we left the EU. Nevertheless, it was a seriously large amount of money, and the UK's Department for Agriculture and Rural Affairs (Defra) calculated that the subsidy payments accounted for more than half of British farms' incomes – around 55 per cent, in fact. There were different schemes available to British farmers, but the basic – so-called Pillar 1 – payments were based purely on the area of land owned. In 2020, the last year under which the EU rules applied, as long as a farmer held at least five hectares (12.35 acres), and was actively working the land, they could claim £233 per hectare (£92 per acre) in basic payments. Just before Brexit, there were 85,000 farmers receiving the CAP payments. In addition to Pillar 1, EU member states were also able to fund Pillar 2 payments with the stated aim of fostering the sustainable management of natural resources, combating climate change and encouraging the balanced development of rural economies. Pillar 1 payments accounted for 80 per cent of the UK's CAP budget, the remainder from Pillar 2.

Even with the subsidies, the government calculated that around 16 per cent of UK farmers were making a loss. Without the subsidy, Defra calculated that 42 per cent of the country's farmers would not make a profit. And, startlingly, the CAP system accounts for more than 60 per cent of the profits made by farms. Under the EU CAP scheme, about a third of the subsidies went towards activities that had an environmental benefit, but it was widely acknowledged to be inefficient – with half of the direct payments going to just 11 per cent of farmers, and only 2 per cent of payments to the lowest-earning 20 per cent. One individual, at the very top of the tree, was getting €3 million in EU subsidies per annum.

The paperwork involved in claiming the CAP subsidy was considerable. Now that Britain has left the EU there will, inevitably, be new types of paperwork. Under the proposed new Agriculture Bill, British farmers will lose their basic payments gradually over the next few years, until they are phased out entirely by 2028. By 2027, the payment per acre will be down to around £12 per acre, or £31 per hectare. However, farmers will have other alternatives as the subsidy reduces. Country Stewardship schemes (supposedly simplified) have been brought in to compensate, or reward, farmers for looking after the natural environment on their farms. There are proposals, as yet unclear, to introduce an Environmental Land Management (ELM) system that will pay landowners in what is famously described as 'public money for public goods'. In 2021, the government asserted (through the then Parliamentary Under Secretary of State Victoria Prentis) that farmers and other land managers will have the chance to enter agreements under which they will deliver clean, plentiful water, clean air, thriving plants and wildlife, a reduction in and prevention of environmental hazards, adaptation to and mitigation of climate change, and finally the slightly vague but ambitious qualities of 'beauty, heritage and engagement with the environment'.

By the time the current subsidy system ends in 2028, the government claims it wants to see a renewed agricultural sector which will be producing healthy food, as well as farms which are both profitable and economically sustainable without the need for subsidies. Farming will also be not simply doing no harm to the environment, but actively contributing to environmental goals. These goals include the UK's commitments towards tackling climate change. As stated in *The Path to Sustainable Farming*, produced by Defra in late 2020, the recovery of nature will accompany a 'rediscovery of the art of good farm husbandry'. That may be the hope for the future. Turn it on its head and it seems to me that we couldn't ask for a more damning official verdict on what has been going on within our agricultural systems up until now.

Worryingly, the National Audit Office has expressed concerns over how ready the new system will be before the subsidies run out. The ELM system is in a pilot phase from 2021 to 2024. The NAO points out, fairly obviously, that implementing the new schemes will be complex and 'high risk'. Working out exactly how the farmers will be paid under the ELM system is still somewhat vague. The new payments will reward environmental benefits under three new frameworks: the Sustainable Farming Incentive, Local Nature Recovery and the Landscape Recovery scheme.

The notion of ELMs was first introduced by Michael Gove when he was Environment Secretary in early 2018. He described the plans as a 'once-in-a-generation opportunity to shape the future of English farming and the environment'. In short, the new schemes will reward farmers who save and conserve 'natural capital' – defined simply as the world's stock of natural resources, including water, air, soil and living beings, all of them essential for human survival. The ideas behind the proposals are backed up by a great deal of science, and irrefutable evidence that our natural environment is ailing. Much of the most compelling proof stemmed from findings of the Natural Capital Committee, which ran from 2012 to 2020 and which included some of the foremost environmental and land management experts in the UK.

Under the chairmanship of Oxford University's Professor of Economic Policy, Dieter Helm, part of their job was to help the government devise and implement a 25-year plan for the UK's natural environment. Professor Helm has for some years been an adviser to the British and other European governments, as well as to the European Commission, notably in the drawing up of the Energy Roadmap 2030. He is also passionate about how our developing society affects the natural world. He is very clear that the whole way of thinking about our relationship with nature has to change. He believes that is the only way we can tackle the big environmental challenges of the century. In his words, 'we need to end the apartheid between economic growth and protecting and enhancing the environment.'

The Natural Capital Committee was focused especially on addressing the unsustainable use of our natural assets, and advising on what actions should be taken to improve and protect the nation's natural capital. It sounds like a vague thing, but nothing could be more fundamental to our survival than our planet's stock of natural resources. What the NCC did was convince the government that they need to think about the natural environment in a very targeted, hard-nosed way. 'Natural capital' isn't just a catchphrase, or a bit of jargon, it describes the fundamental resources that humanity depends on to survive; it is provided for free by nature and includes our air, our water, our soil, and the plants we use for medicine, building, food and fuel. It encompasses the forests that protect us from the sun and from flooding, and, crucially, it includes 'ecosystem services', for example, from insects which pollinate our crops, as well as the vegetation and soils that we know store vast amounts of carbon.

Natural capital assets extend in perpetuity, and have a special, open-ended value extending into the far-distant future – as long as they are not wasted. However, some are renewable (like fresh water) and some are not (like oil and gas). For the last century and more – perhaps since the Industrial Revolution – humanity has undervalued these things, or assumed they are in inexhaustible supply. The NCC was very much focused on advising the government on what it needed to do to put the economy on a sustainable footing regarding the environment. Like any asset, natural capital can be squandered but, unlike most assets, it is often irreplaceable once it has been used up. As Mark Twain said, 'Buy land – they aren't making any more of it.'

Professor Helm is starkly realistic in his thinking. He accepts that not everything can be preserved, and points out that those parts of our natural capital which have been damaged can be compensated for by making gains in other areas. That seems to me to describe the microcosm of a farm very well, where we sacrifice certain parts of so-called 'wild nature' in order to produce food, but we boost and create other areas on the same land in ways which foster growth and abundance – natural capital. One of his important contributions to changing the way we think about farming is to ask us all to imagine its future. He argues, wisely, that none of us alive today can predict what the world will look like in thirty or forty years' time. As he says, when he was writing his thesis at university on a typewriter he could not have imagined that he would one day be using a laptop computer or, more recently, that he would be able to run much of his life with a mobile telephone. Technology, he says, has a way of upsetting what we think we know about how the world works. And for farmers he has a warning – don't assume that in the future we will necessarily need the land we currently use to produce food. We might need more of it for housing – an extremely likely scenario. But we might be producing more food with things like 'vertical farms' where crops grow up a wall in climate-controlled conditions. Test sites have shown that these can be many times more productive than the same area of traditional 'horizontal' planting out in the open. Some commercial vegetable production is already being done using hydroponics, for example, which can successfully grow things like lettuce, tomatoes and cucumbers using much less water than traditional methods. Meanwhile, several companies are working towards large-scale insect production, with several species known to be very rich in amino acids and with high protein-to-weight ratios. Mealworms may well be the future for our diet.

However, while we wait hopefully for the food revolution and new sources of protein that can be produced more sustainably, we have to deal with the realities of how we produce food right now under the current system. I'm not vegan or vegetarian, but it seems pretty clear that for the health of the planet we need to be eating less meat, and certainly less cheap meat produced overseas using unsustainable, high-carbon methods. To meet our reductions in greenhouse gas emissions we are already being told to eat less animal protein. And there are plenty of health experts who say too much meat is bad for us as individuals. Unsurprisingly, for both health and emissions reasons, the government's Climate Change Committee recommends that everyone reduces their intake of beef, lamb and dairy by 20 per cent, which along with a commitment to reduce food wastage by 20 per cent would save the UK 7 million tonnes of CO_2 emissions by 2050.

It's no good reducing our own carbon footprint from livestock farming if we simply replace it with meat shipped in from other countries, especially if those countries aren't taking measures to reduce their own emissions. The boffins in the civil service call this 'carbon leakage from trade in agricultural products'. And, even if you're sceptical about switching to a more plant-based diet, it's entirely logical to think we should be encouraging people in the UK to eat British animal protein, which has the major benefit of not being shipped from halfway across the planet. And we need to accept that we should pay a fair market price for that meat, allowing the farmer to produce it to high welfare standards and with the best environmental credentials. It's why a post-Brexit Britain cannot betray the farmers by allowing food imports to out-compete them. If we are serious about reducing carbon emissions, then such 'cheap' imports would need to have their carbon production and emissions costed into their price.

It's worth pointing out that while we all know by now that meat production is one of the major drivers behind deforestation in the Amazon basin and other parts of the tropics, the environmental consequences of raising cattle in this country are much less damaging. For a start, we don't have rainforest – which currently stores about a quarter of all the carbon on the planet's land. Britain's temperate climate is much better suited to growing grass, and more than 40 per cent of our land is covered with pasture and semi-natural grassland. That's about 450,000 hectares, or just over 1.1 million acres, which is more than 350,000 times the size of Trafalgar Square, or about a fifth the size of Wales. That grassland is not particularly suited to growing arable crops, but it *is* good for grazing, and almost 90 per cent of British beef is produced on predominantly forage-based diets. The presence of grazing livestock does not mean that all other species and plants are necessarily excluded from the land, or that it does not provide wildlife habitat. The Committee on Climate Change acknowledged that emissions from beef production in the UK were around half the global average. Soya, another plant now regarded as the scourge of the rainforest, plays little part in animal feed in the UK.

Attaching value to things that seem to be supplied 'for free' by nature, isn't straightforward under our capitalist system. In an attempt to quantify the value of 'nature', the government published a review into *The Economics of Biodiversity* in early 2021. Led by Emeritus Professor Sir Partha Dasgupta at Cambridge University, it acknowledged that nature is much more than a 'mere economic good', and made the distinction between how things may be 'valued' in an economy, but without taking into account their 'intrinsic worth'. In nature, said the authors, humanity must think of itself as a natural capital asset manager, but those assets are not under our individual control: their safety and value are under the control of all of the people in the world, whether we are farmers or fishermen, households or companies, governments or communities. One of the major problems with trying to account for all of nature's assets is that they move – they actually crawl, run, swim and fly. And nature has other difficult qualities when it comes to assessing their economic value – many of its processes are silent and invisible – like what is happening beneath the soil much of the time. And often we don't know whom exactly should be held responsible when human actions affect those processes.

Only recently I was asked by one of our farmers what to do about a three-acre field where he had ten lapwings nesting. He knew that lapwings had been declining in England, and that fewer and fewer of the birds nesting here had been successful at rearing their broods. This small example illustrates one of the big problems with natural capital: we don't know how to value things, or, crudely, what sort of price to put on them. In a capitalist economy where everything is measured in terms of growth and appreciation, that's a challenge. As Dieter Helm points out, nature gives us this immensely valuable set of resources, for free.

Now, take the case of the farmer and his lapwings: it would cost approximately £1,500 per acre to plant the three-acre field with potatoes (depending on the variety used). That would probably yield between 25 and 30 tonnes of potatoes per acre, worth anything from £100 to £400 per tonne, depending on the season. That gives the farmer a return of between £2,500 and £14,000. The ten lapwings might reasonably be valued therefore at £250 per bird, or as much as £1,400. Some people would consider that, at the lower end of that scale, the birds would be relatively cheap, and would be willing to raise the money to pay for them rather than let the farmer plough, sow and harvest his potatoes. But if the government was going to compensate that farmer for not disturbing the birds, while that money could otherwise be used for something like the NHS, then how many people would think it a good deal? In the real world, those are the kinds of hard choices we may have to make.

It seems to me that waiting for the government – indeed any government – to fix the natural world is a flawed wish. The very thorough 2021 Dasgupta Review runs to more than 500 pages and is a heavily academic examination of hard economics and nature. But to me it reaches a chilling and very simple conclusion: it reiterates that 'the consequences of actions which desecrate Nature are often untraceable to those responsible.

Neither the rule of law, nor the dictates of social norms are sufficient to make us account for Nature. We will have to rely on self-enforcement, to be our own judge and jury. And that cannot happen unless we create an environment in which, from an early age, we are able to connect with Nature.' It really is up to each and every one of us to do what we can to fix things. But we won't do it unless we care. And we can't care about the natural world if we don't see it first-hand, and if we simply don't understand it on the most basic level. In our increasingly urbanised societies, it is very easy for children to grow up thinking that the natural world is somewhere else, somewhere 'out there'. As long as there is electricity to power our computers, food on the supermarket shelves, something distracting to watch on our screens, the reality of the natural world is all too often a mere concept. It's not surprising when you consider that almost 85 per cent of the UK population is now urban.

It's no good just saying that it's difficult to put a price on our natural resources. We know that already. But we also already know that without a healthy ecosystem, people cannot survive. It sometimes feels like as long as there is still food in the supermarkets, and petrol for our cars, not much is going to change. Because in wealthy countries like ours, most people don't have to face the grim truth about what is happening to the natural world. The statistics and the danger signs are notional, a passing word on the news or an appeal on social media on behalf of an endangered species. For a time it looked as if the Covid-19 pandemic was making people wake up to the idea that something unexpected could come along and disrupt everything we thought was safe and secure. There were stories along the lines of 'this is what happens when we tamper with nature'. And there were calls for a 'great reset' that would involve less consumerism and fewer flights abroad. But that effect seems largely to have worn off. And, in spite of the frightening glimpse the pandemic gave us of how fragile some of the things we do are and how easily threatened some of our habits may be – like going abroad whenever we feel like it for a holiday – the amazing thing is that, by and large, no one starved due to a failure in the food-distribution chain. Strangely, for many people stockpiling loo paper seemed to be their highest priority. Thankfully, the developed world's food supply chains held up remarkably well.

If everything is connected in an ecosystem, then how do we go about prioritising which species we decide are worth saving? The countryside is not a zoo: we can't simply fence off the odd field and say it's reserved for wildlife. We know that many of those basic elements in our landscape that we may label 'natural capital' have already been badly compromised. Ever since the passing of the 1947 Agriculture Act, (a perfectly understandable attempt to make farming and food production a modern industry that would support economic recovery after the Second World War) the land has sickened. The global conflict had shown us how potentially fragile our island nation could be when it was largely cut off from the rest of the world. But, even today, the symptoms of how impoverished our natural world is, and what it gives us, are there if we look for them: the declines in biodiversity are no longer 'what ifs?' Through what Dieter Helm calls the 'relentless march of agri-chemistry and farm mechanisation', our soils have become depleted, and many of the archetypal species which we think of as part and parcel of the English countryside have all but disappeared.

We've read it so many times – we have lost 97 per cent of our wildflower meadows in Britain, and suffered severe population declines in half of our farmland birds. Using data from more than seventy organisations, the most recent *UK State of Nature* report monitored almost seven hundred species of mammals, birds, butterflies and moths to show that 41 per cent of them had shown moderate or strong decreases in abundance over the last fifty years. Some species are now doing better than they have in the past, but over the last decade 53 per cent of UK species are showing what the report calls 'strong changes in abundance'. When it comes to birds, the statistics are perhaps most shocking, with an estimated loss of 44 million individual birds from the countryside compared with fifty years ago. You can begin to understand why I feel a sense of urgency about reversing some of these losses.

It's hard to comprehend what's going on with the natural environment when we are presented with these big numbers and statistics. In general, it's easier to start understanding things when we focus on animals or birds that we think of as part of our world, and in my case that means an English landscape. It helps if we look at some of the specific species that we all recognise. Take a common, but stunningly pretty bird like the starling (*Sturnus vulgaris*). It's easy to miss, in flight quite small and dark against the sky, just about the size of a blackbird – but if you get close enough you'll see it has a bright yellow bill. In their winter plumage they are flecked with white spots (actually the tips of newly emerging feathers), but in the spring and summer when they are breeding, they develop an incredible iridescence within their feathers so that they shine and glisten, like the purple and green when a spot of petrol has been dropped into a puddle of rainwater. The English starling (or European variety) was introduced into Central Park in New York in 1890 by a man called Eugene Schiefflin, who, the story goes, wanted all of the birds mentioned by Shakespeare to be seen in the USA. The experiment was successful, because there are now about 200 million starlings in America. Back in Europe, the bird that creates magical shapes in the sky called murmurations when it gathers in winter flocks thousands strong, is in trouble. When I worked in the West End of London (after abandoning school), there were always huge numbers of starlings roosting in Leicester Square, at one time an estimated 100,000 of them. Perhaps they didn't have the same romantic appeal as the proverbial nightingale in Berkeley Square, but when the winter sky got really dark in November they filled the night with the most incredible sound, a rippling wall of gentle cheeping in the branches of the tiny park in the middle of all the cinemas and fast food restaurants. For me, and I'm sure for many other Londoners, their busy chatter was a comforting reminder that there were creatures out there that lived by a natural rhythm, even though the birds had found one of the busiest spots in the city to come together. Those starlings have all gone now – possibly driven out because they were seen as 'dirty'. Overall, in this country these birds have seen a steady decline since the early 1980s, almost certainly due to a lack of earthworms and other insects in the countryside. Whatever the reason, the British Trust for Ornithology (BTO) says numbers are down by almost 70 per cent. Another bird, the turtle dove (*Streptopelia turtur*) is actually the bird mentioned the most by Shakespeare – 66 times in 22 plays. Its populations are in an even sorrier state. I remember them visiting Raveningham. Each spring we would see one particular pair nesting in a dead pine tree. In about 2013 I noticed that there was just one bird, calling for its mate. When the dead pine was removed, the dove moved to a nearby copse where several ash trees grew near to a pond. For three or perhaps four years, the lone turtle dove waited each spring for its mate, until one year it didn't return. At about the same time, the stand of ash trees succumbed to ash dieback. The demise of the doves and the trees were no longer things I heard mentioned on the radio, or read about in the news. They were happening in front of my eyes.

In winter, turtle doves stay in West Africa, coming to Europe – like cuckoos – to breed during our summer. The RSPB estimates their numbers to be 93 per cent below the levels of the 1970s. These pretty blush-pink birds, with strikingly rich mottled wings and delicate bars of black and white feathers on the nape of the neck, are the UK's only migratory dove. Part of the turtle doves' vulnerability is that they eat grains and seeds almost exclusively. With the 'cleaning' of crops with herbicides, the quantity of wild arable plants like fumitory, chickweed and cereals with suitable seeds available for the birds has been massively reduced. Once upon a time they were able to feed on the grains left over in the farmer's winter stubble – the stalks left in the field after harvesting – something that isn't often an option today. Turtle doves also favour tall dense hedges in which to nest, often choosing impenetrable scrubby growth like hawthorn. Many ornithologists are now pessimistic, bracing themselves for the very real possibility of the global extinction of turtle doves, but long before then, their loss as a breeding species from the English countryside seems almost inevitable. It's a tragic fate for the bird, whose purring call was so brilliantly described by Isabella Tree as a 'fading pulse from the landscape of the Elizabethans'.

The list of potential losses seems to go on and on. It's not just turtle doves and starlings, but also kittiwakes and corn buntings, skylarks and lapwings that have suffered in recent decades and begun a seemingly terminal decline. We have lost most of our hay meadows and a huge percentage of lowland ponds. Meanwhile, diseases like ash dieback and oak decline are doing great damage to some of the trees that we think of as emblems of the English countryside. There are, of course, some glimmers of hope for a few individual species, such as the return of otters from the brink of extinction after pesticides killed off so many of their river habitats in the 1960s and 1970s. There has been the recovery of birds like the red kite, and rises in numbers of goldfinches, to name some of the bright spots, albeit much fewer in number. The *State of Nature* census covers 7,615 species, of which a staggering 13 per cent (971) are actually at risk of imminent extinction. The data also show that changes in the climate are contributing to significant changes in distribution, with many species of birds, insects and plants moving into areas where they have not been seen previously.

There are numerous complex factors that account for this decline in overall biodiversity. They include climate change, the loss of traditional mixed farming and a consequent loss of habitat diversity as monoculture crops begin to dominate the land. The greater use of pesticides has reduced the number of insects, a crucial food source not just for many birds, but also for other invertebrates. One of the biggest changes in British farming in recent decades has been the increasing prevalence of winter crops, developed so that they could be sown in the autumn. Farmers favour them because they are more reliable than things sown in the spring, where temperatures and rainfall have tended to be much less consistent, especially in recent years. But traditionally, after the autumn harvest the fields of stubble would have been left fallow, providing a food resource for over-wintering birds. Ploughing would have taken place slowly and gradually over the winter months ready for spring sowing. Now, modern equipment means large areas can be ploughed within days and then sprayed to keep weeds away, or covered in manure to prepare it for the next spring's sowing. All too often, the result is that there is nothing left in the fields for the birds to eat in the winter. And that relatively open stubble habitat provided not just food, but also crucial nesting areas for species like lapwings, skylarks, corn buntings and curlews. Meanwhile, improved drainage has led to the disappearance of the damp, boggy areas where birds would have once found food in the form of plants and invertebrates. It's clear that if there are remedies for this, then they lie in the hands of farmers and landowners.

I don't find these doom-laden studies about our impoverished natural world a reason to lose hope. The vast majority of the species under threat spend a significant part of their lives on our farmland. That can be seen as a massive positive opportunity, rather than a death sentence. Under the newly proposed government plans, the definition of what may classify as a 'public good' is wide. It includes many things which enhance our natural environment at the same time as they allow for more sustainable food production. Public goods would include things like investments in new technology that increase productivity, as well as less tangible actions including providing better public access to farmland and the countryside. Farmers or landowners who make changes to the way they operate that may mitigate the effects of climate change will also be rewarded, as will those who can show that they have increased biodiversity on their farms, contributed to better water quality or raised welfare standards for their livestock. It all sounds quite radical, but the idea behind it all is simple: the old system – based on subsidies paid out purely according to the amount of land you farmed – wasn't protecting or enhancing our natural environment. Indeed, in many cases it was actively harming the countryside and all that lives within it. It didn't persuade farmers – who control what happens to about three-quarters of the land in the UK – to create a habitat rich in wildlife or in any way resilient to climate change, or even the entirely natural fluctuations in the weather.

I am a countryman, but I am neither a landowner nor a farmer. And yet, since these radical changes were first proposed to the systems which have supported farming for decades, I have had a constant stream of people calling me at Holkham to ask for my advice on how to make their land management more positive for the environment. Those looking for guidance, or ideas, have included ordinary tenant farmers whose holdings amount to a hundred hectares or even less, all the way up to some of the best-known landowners in the country, some of them scions of historic estates with many thousands of hectares.

Part of what I was hired to do at Holkham was to continually enhance natural capital. And yet I don't have formal qualifications in agricultural management, nor do I have a degree in environmental science. I've learned what I know by going out into the fields on a daily basis since I was eighteen. In summer I get up at four a.m. and even in midwinter I'll be up before six. Much of what I want to talk about in this book stems from what I'm doing now in north Norfolk. But before I came to Holkham, I worked for more than twenty years at another old family estate in the same county, and before that on other estates in southern England. My knowledge has been gained in very practical ways and not so much from studying books. I have always learned from others, whether they be scientists or farmers, gamekeepers or naturalists.

The vision I have for restoring the balance between farming and nature is not restricted to this patch of Norfolk. I want to get other parts of the country to do the same sorts of things – because only if we can make these positive changes on a whole-landscape scale will we save the countryside. It's no good having one or two cases of good natural habitat surrounded by an ecological desert. Because it's all connected, you see. Birds and insects, mammals too, move around. That may seem pretty obvious, but it seems like we've forgotten some very obvious things, along with some other simple truths.

3
In the Limelight

I've got one of those long family names that makes people assume things. It's a name that a lot of people have heard of, and because of the way English society works, they think I must be a 'toff'. In fact, in my childhood my parents worried constantly about money, and we moved frequently, scarcely putting down roots anywhere. I'm one of six children, and one half of the set of twins born at Odstock Hospital near Salisbury in 1970. My mother, Jini (Jennifer) Lash, was a writer. My father, Mark, left school early due to ill health, and tried many things in his life, including working on ranches with sheep and cattle in Texas and Australia. When he returned to England he took on the tenancy of a mixed farm on land that belonged to the Earl of Stradbroke, near Southwold in Suffolk. He was farming when he met Jini, and they stayed there until just before my brother and I were born. My father would eventually become a respected photographer, and much of his work involved taking portraits of some of England's finest country houses.

I have well-known relatives, especially some of my

siblings, who have made a success of careers in the media and entertainment worlds. Our long name comes courtesy of being great, great, great grandchildren of Frederick Twistelton-Wykeham-Fiennes, the 16th Baron Saye and Sele. Our cousins, distant and near, include the explorer Sir Ranulph Twistelton-Wykeham-Fiennes, and the writer William Fiennes. My eldest brother, Ralph, is pretty well known for starring in some very successful films, including *Schindler's List*, *The English Patient* and, of course, the Harry Potter and James Bond blockbusters. He also produces and directs, and is an accomplished stage actor. My brother Magnus is a composer, and my 'younger' twin, Joseph, is also a successful actor with numerous successes on the big screen and television. My sister Sophie is also a film-maker and documentarist. Joe and I are the youngest of the six siblings and we also have an older foster brother, Michael. My mother brought Michael into the family after seeing an appeal on his behalf in a local newspaper for a boy with a 'desperate need for a home where he can read a book'.

How I came to do what I do now is what you could call a complicated story. Making my life in conservation was never planned, nor was it achieved through a straight or conventional route. Perhaps, though, the seeds of it were sown all along the way, starting in my early childhood. I suppose I had what some people would think of as an unusual, or bohemian upbringing. As a child, whatever your family life is like you take as normal. Like many families of the time, we had ancestors who had done service in the colonies and the military. My paternal grandfather was in the steel industry and my mother was born in India (her father, Henry, had been in the Indian Army), and the family came back to England in 1948, after Partition. Mother was ten years old when they came home, but said she didn't even know how to run a bath. She had to learn. I did meet my grandfather a few times, and I remember being very excited because he had a tiger-skin rug in his house, made from an animal he had shot himself. His brother also had a career in the Raj, and was for many years the Bishop of Bombay. I have a dim recollection of him baptising me when I was about three in our garden in Dorset. I'm convinced I was naked except for a pair of wellies, but no one else is around to confirm it. It was soon afterwards that my parents decided to try living in Ireland and bought a townhouse in Kilkenny – as an investment project because life in England had got very expensive. I remember the house being very tall, and having what seemed to me to be a very large garden. It also had a big copper bath with great shining pipes and taps and levers that controlled the temperature. I think my father felt that life in Ireland would be less frenetic than in England; and perhaps as a writer it appealed to my mother more too. Her own mother was Irish, and I expect that played a part. It wasn't really as romantic as it might sound. There was never any spare cash, and they had to call on the help of a lot of friends.

Eventually, my parents decided that buying houses in various states of disrepair and doing them up was the only way to make a decent income. It meant that we moved home more than a dozen times and, in all, I went to thirteen schools. Not ideal. I remember my younger years in the west of Ireland perhaps best of all, those we spent in the small house close to Sheep's Head on Dunmanus Bay in County Cork. It was a wild and dramatic place, and living there was like being part of the natural landscape. My parents built the house, or at least restored it, and I remember it then as an ugly thing, but it stood in a spectacular setting. We certainly didn't have money. As the youngest children, Joe and I wore hand-me-downs, and never had any new clothes. Even our children's storybooks were battered and worn and had been in the family for years.

For me, the big memory from those years in Ireland was that I spent a lot of time in hospital. The doctors were never exactly sure what was wrong, but when I was about four I became more and more reluctant to walk. They suspected some kind of bone infection, and the treatment was to be put in plaster and kept in traction for several months. The hospital was several hours' drive away on slow, winding country roads. My mother would visit when she could, but I was alone for a lot of the time. When the cast eventually came off, the first thing I vividly remember was being allowed to go rock-pooling with the other kids. That life had its charm, but it was all very chaotic in a way, and I didn't go to a proper school until I was seven.

I have mixed-up memories of lots of different houses, and after Ireland we moved back to Wiltshire, where for a time we all stayed in a flat that was in a farmyard near the ruins of Wardour Castle. There was a battery chicken farm nearby and I remember seeing lots of dead birds and a cart being loaded up with the carcases of hundreds of chicks. There were dark woods, and lots of playing outside and around the farm. At the next house, not far away in the small town of Tisbury, we had a garden big enough to keep some caged rabbits. They were for eating, rather than pets, and I clearly recall me and my brother pulling the skins off the carcases in that bloodthirsty way that small boys enjoy. The local newsagent called around once to deliver something and fainted when he came into the kitchen to find us covered in blood and guts and fur all around. I can also remember being in the garden once and a blue tit coming and perching on my hand as I scattered some birdseed.

These things stand out from my young childhood, and I also have strong memories of my father reading us bedtime stories. There was an Edwardian novel called *Knock Three Times!* which involved what I remember as a deranged pumpkin and was completely terrifying. My father was tall and good-looking; to me he looked a bit like James Bond. He spent a lot of time shut up in his darkroom with his photography, and would emerge at supper time to join us. I was a little scared of him because he would beat us with a slipper if we were naughty, although that was pretty standard then. I remember we almost never went anywhere on holiday because there was no money, but once my father took me and Joe with him when he was loaned a cottage in Wales for a few days. He said it was time we learned to fish and took us down to the beach. It was freezing. He kept shouting at me to go further out from the shore, and I had to keep going until I was almost up to my waist so that I could cast the line into deeper water. It was bitterly cold and I started crying with the discomfort.

He was, I suppose, quite a manly man, but he was also a very domesticated father really. When he arrived back in Wiltshire from London, he'd often find a pile of ironing waiting to be done, because my mother spent extended periods lying down, suffering with recurring slipped discs that made standing for any length of time difficult. I don't have memories of spending a lot of what they now call 'quality time' with him. And with my mother being ill, I spent a lot of time playing outside on my own, building dens, trying to catch animals and insects, that kind of thing, which was seen as nothing strange or unusual for boys back then. I was quite happy.

We moved yet again to Salisbury when I was about seven, and then I remember being sent to a private prep school where I had to wear a uniform of shorts, brown sandals, blazer and a cap. There were wooden desks with inkwells, and we had to learn to use a fountain pen – no biros allowed. We didn't stay long because our parents moved back to London, to what my mother called Wandsworth and my father called Putney. Joe and I were sent to the local primary wearing our old school uniforms from Wiltshire and of course immediately got picked upon, because everyone else was in civvies. We were told to deal with it, as there was no spare cash for new clothes. We weren't there long before mother decided she wanted to be back in the country. Joe and I and our sister Sophie ended up near Shaftesbury in Dorset, while our older siblings stayed in London with our father, although he used to come and visit us often at weekends. In a way, the family functioned as two halves, even though their relationship remained close, and strong.

A couple of schools later, we were briefly at a private boarding school where Joe was a prefect and I, unsurprisingly, was not. It didn't bother me, and I think I convinced myself that being a prefect wasn't cool anyway. Eventually, I got a fairly average selection of O levels and an 'Ungraded' mark in CSE Maths. It probably didn't help that I had been smoking dope and was stoned when I turned up for some of the exams. But, by some miracle, I did all right in English, Art, Technical Drawing and Metalwork. Even so, I knew by then that school wasn't for me, and decided to leave. Joe was put into a different class, and we didn't mix at school much. We're not identical twins and we've always had our own lives and separate friends. He was always more academic, although he didn't find it came naturally and I remember my mother sitting with him for hours, patiently teaching him to read. For me it was much harder, and when I was twelve she took me to see someone in Marlborough where they tested me and confirmed that I was severely dyslexic. It wasn't any kind of momentous shock. All I remember about that day was that my mother and I went to the famous old Polly Tearooms afterwards – a real treat.

After scraping through my O levels, I knew I didn't want to stay at school, but I started at a sixth-form college. I was beginning to get interested in girls, of course, and soon contracted glandular fever, which they used to call 'the kissing disease'. That kept me off school for a couple of months, and I spent less and less time on my studies, and after returning to class decided within a term that I wanted to go to London. I stayed with my father for a while in his small flat in Pimlico, where I slept on the sofa bed until I found a job. It's not much of an exaggeration to say I soon fell into fast living. A friend of a friend said they were looking for someone to work at a nightclub in the West End. It was the Limelight, a former Presbyterian chapel on Shaftesbury Avenue that had been converted into one of the coolest party places in town. I had been into music for a while – especially Bronx Hip Hop – and in the last couple of years at school had worked with a mobile disco, another factor that didn't help with my studies. My mother was relaxed about it all, but I'd get home in the early hours of a Sunday morning and often find her dozing in a rocking chair having waited up for me. She didn't discourage me at all.

The Limelight was part of a chain of clubs originally started in Atlanta and New York by a colourful man called Peter Gatien. He was a Canadian entrepreneur and club promoter who always seemed to me a larger-than-life character. He wore an eyepatch and had all the swagger of a pirate. I worked as a receptionist and general dogsbody at the club, but soon got on first-name terms with a lot of the big stars who passed through the doors. Bob Geldof, Boy George and George Michael were regulars, and in the VIP area you'd get used to seeing Tina Turner dancing on a table after a concert, or Fleetwood Mac chatting to Cher. A typical night would find me in the VIP area with some of the most successful stars of the time, people like LL Cool J, Lemmie, Rick Astley and Bros. It was an exciting place for a teenager, and the music pumped out through a revolutionary quadraphonic speaker system so that every night was like being at a live concert. It was an intoxicating environment, in every sense of the word. I lived at night and drank a lot of Stolichnaya on ice. And of course there was a lot of dope and coke around in those kinds of circles. I remember the cleaner who came in the morning was always high as a kite because he would find enough cocaine dust left in the loos to keep him happy. It's fair to say I sampled a lifestyle that was at best precarious and at worst seriously unhealthy. I vaguely recall a period of three months when I didn't really sleep at all. It was a far cry from village life in Wiltshire.

I stayed at the Limelight for a couple of years, and then I started to get ill. I developed severe eczema and generally got very run down. Although I was barely nineteen, I somehow worked out that the night-owl existence and the hard partying were probably to blame. By then I'd seen enough of what could happen to some of the people who passed through the doors of the Limelight to understand that the long-term prospects weren't bright if I stayed. Even at that age, you know that when people start offering you heroin regularly it's probably time to leave. I didn't have anywhere to go, but in the school holidays before I went to London I had done some work on the Knepp Estate in Sussex, helping out with lambing. Somehow, I had enough sense to ask the owner, Charlie Burrell, if I could come and stay for a few days while I considered my future. He generously agreed that a change of scene might be a good idea.

Today, Knepp is famous for its ambitious 'rewilding' programme, and the estate is now a test case for what can be done to restore healthy ecosystems on land that has been farmed intensively for decades. Back then, it was still much more of a traditional working farm, and there was plenty of hard physical work for me to pitch in on with the animals and the land. Although I had arrived rather jaded from my nocturnal life in the West End, I wasn't totally ignorant about the countryside, because many of the numerous houses we had lived in had been in rural areas. We always had animals – dogs, of course, but also rabbits and chickens that were for the pot rather than pets. My mother was always adopting dogs, rescuing them and giving them a proper, loving home. It almost always worked out, except for one lurcher I remember. He bit the postman and had to go. Rural life certainly wasn't a sanitised thing. We had several lurchers, and I recall going to a sort of gypsy fair once with mother and watching a dog give birth to puppies. They kept the bitches but the male pups were killed with a hammer on an anvil in front of us. I think you accept those things when you see them as a child. I remember a fox getting into our chicken coop once and finding the carnage of all the dead animals the next morning. I wasn't upset particularly, it was just something that happened. At the time I didn't have any sort of natural attraction or ambition to get into farming.

My mother liked the natural world, and although we didn't have the money to travel around very much, she once took me and Joe to Selborne in Hampshire, to see Gilbert White's house. It wasn't very far away from where we lived, and perhaps something of the atmosphere of the place, or some half-remembered passage of his natural history, lodged in my mind. He was certainly an example of a naturalist who believed in first-hand observation, and was never shy of spending time on the small things, which he knew were connected to bigger things. Charles Darwin credited White's 1789 *Natural History of Selborne* as a key influence. White's description of earthworms is classic: 'Though in appearance a small and despicable link in the chain of nature, yet, if lost, they would make a lamentable chasm.' I love the idea that even in the eighteenth century someone could make that link between the soil and then on to the worms at the base of the pyramid of life that depends on it.

Our childhood wasn't exactly *Swallows and Amazons*, but there was a good deal of outside activity, and I would catch beetles, fireflies and slow-worms to keep as pets. Like many boys, I did enjoy discovering dead animals in the countryside. It was a chance to really look at what they were like up close, rather than relying on pictures in books. I still have the handsome young dog fox that I found dead on the road, and which my mother had stuffed and mounted for my sixteenth birthday present. It's very well done (by Howard 'Benny' Bennet, the master taxidermist, who is, coincidentally, a Norfolk man), and extremely lifelike as it stands up on its hind legs in the act of pouncing on a weasel emerging from a tree trunk – another dead animal we found the same day. For a time I had a very nice stuffed stoat, too, until the dog ate it. I remember that my mother befriended a local character whom everyone called 'the Romany King', and he had something to do with a museum of gypsy life near to where we lived in Wiltshire. He showed me how to catch goldfinches in a simple home-made trap – birds he could sell at the market in those days as caged pets. Of course, I tried catching birds myself, and was successful with a blue tit, using a simple box propped up with a stick to which I had tied a length of cotton. I put some bird seeds around the box with a little trail leading underneath. Once the tit was inside, I pulled the string so that the box fell and trapped it. Then I kept the bird prisoner by tying a length of cotton to its leg, so it could still hop around and fly a little bit. When my mother discovered what I'd done she was furious and made me set it free. I didn't try the other skill he showed me – how to cook hedgehogs by baking them in clay in a fire. She didn't object to me catching wasps, moths and other flying insects. Once, a large hornet that I had fixed to a display board suddenly came to life during a Sunday lunch and flew around the dining room with the pin and its label still attached. I thought I had killed it by putting it into the freezer for a day, but the warm room revived it. My mother thought it was hilarious. And of course, like a lot of boys then, I tried my hand at taxidermy, probably inspired by a copy of Durrell's *The Amateur Naturalist*, which describes a lot of that sort of thing. The first attempt was with a mouse, but it was far too small and fiddly and I soon gave up. I had to make do with catching a mink and pinning the skin to a board as a trophy.

I suppose I absorbed a lot of country lore through my father, who was a passionate grower of vegetables, even after he gave up farming professionally. My mother took me on long walks, and I know she made me notice things about the natural world, especially the changing seasons. I remember a long walk once when, after hours of tramping through the countryside and along a path that skirted a lake, we got back to the car to discover that she couldn't find her keys. We were stranded, and of course this was years before the advent of mobile phones, so it was quite an inconvenience. I can't have been more than eight or nine, but I suddenly realised that we had stopped at a particular bush for a long time, admiring the burgeoning fruits that were developing. 'I know where the keys will be!' I exclaimed, and ran off, retracing our route and finding the spot on the bridle path where we had seen the fruiting trees. Sure enough, the keys were somewhere there on the ground. I remember it now clearly, as in my mind's eye the walk had been punctuated by all of the things my mother had made me stop to observe and really take notice. She taught me to notice the smallest changes in the natural landscape and think about what they meant. Perhaps it's because I don't focus on formal ways of learning to remember things and prefer visual clues, but it's a habit that stays with me.

I've sometimes been asked why I didn't do something 'creative' like my siblings. For a long time I just thought it was a case of having different interests. They aren't particularly keen on nature and I had no ambitions to be a performer. More recently I've started to think that what I do isn't so very different from their world. In effect, I'm creating a show of my own. I suppose it's my version of 'All the world's a stage,' and the players just happen to be birds and insects, plants and water. All of these things have their parts to play in the natural system, which is more complex than any man-made world. Being Director of Conservation requires a lot of piecing together of very disparate and volatile elements to create a functioning whole.

To this day, Charlie remembers me turning up at Knepp from London in my trendy gear – a long black leather coat and rather prissy gloves which only lasted a couple of days on the farm. I ended up staying in the big house with Charlie for almost three years, and after helping out in general on the estate was employed, while still a houseguest, as an all-round helper and labourer. It was a complete change from my London adventures, and I went from earning almost £30,000 a year and having an expense account to earning about £100 a week, although to be fair to Charlie, I also got my board and lodging. I worked in the plant nursery and with the woodsman – a wonderful man called Chris Wagstaff. He was passionate about trees and how to look after them, and he would take the time to explain things like the importance of the hazel stands within the oak forest and how they provide habitat and food for so many other species. Coppiced hazel lives for more than a hundred years, sometimes longer, and is shelter for so many ground nesting birds, including nightingales and nightjars. Hazelnuts are critical for the survival of dormice, but many of our most charismatic birds eat them too: jays, nuthatches and woodpeckers, to name a few. Chris taught me how to cleft chestnut stakes with splitting wedges, a satisfying job that gave a much more rustic-looking result than using power tools.

The jobs I did at Knepp weren't remotely glamorous. I started off building raised beds for plants, planting seedlings and propagating plants that would be sold on to garden centres. I learned how to use a chainsaw and drag timber around behind a tractor. It was also where I taught myself to shoot properly, first with a rifle and then a shotgun. I'd like to say that my father taught me to shoot, but he was pretty hopeless at it. Charlie let me watch everything that was happening, whether it was hearing him dealing with tenancy agreements in the estate office or looking for new ways to make money from the farm. I began to understand how little money was to be made from sheep, how the estate had to hold things like steam rallies to bring in people, and learned that there was more profit to be made from arable crops that would be cut with a combine harvester. I also worked in the game department, and went out 'lamping' – using a high-powered torch and a rifle to shoot foxes. There were all-night sessions keeping watch and babysitting the young pheasant poults. I learned a lot at Knepp, and even made ice cream and yoghurt in the dairy. Most importantly, I gained a real understanding of the different elements of the working countryside, and how what happens on one part of the land affects the others. Knepp has since become famous thanks in large part to *Wilding*, the bestselling book written by Isabella Tree. She was Charlie's girlfriend when I was there, and later married him. In fact, it was after they got back from their honeymoon in Papua New Guinea that I felt it was time to move out of the big house and leave them in peace.

I had vaguely thought about doing some work abroad and, since my father had worked on cattle ranches in Australia in his youth, I was tempted to give it a try, but I didn't have the money to just set off and didn't think my plans would ever come to anything. Perhaps I hadn't quite given up some of the risk-taking habits I had developed in London, because one day I got a tip-off that there was a good bet to be had on the 1990 Grand National. The horse was called Mr Frisk (ridden by Marcus Armytage), and I scraped together £100 to make the bet. Luckily, he set a new course record at Aintree, winning the National at the generous odds of 16-1. Thanks to him, I suddenly had the money to get to Australia.

That was another learning experience, and, through

friends of friends in England who knew people who were farmers, I became the 'Pommie Jackaroo', mustering sheep and cattle on a massive stretch of Outback in Western Australia. I was very green in many ways, and I remember taking a very long bus ride out into the Pilbara, an enormous and pretty unpopulated area of desert in the north-west. I had to tell the driver I wanted to stop at Onslow Turning and he let me out in a vast empty area of scrub with not a building in sight. As the bus drove off in a cloud of dust I wondered if I was going to be stranded and die of thirst. After a few minutes I heard an engine, and then a small plane descended from the bright blue sky and landed near the road. 'Are you Jake?' said the cheerful Aussie pilot as he climbed out. Coming from England, I couldn't really comprehend how big the Outback was. In the place where I ended up working, I think there were only about 50,000 people living in an area of about 200,000 square miles – pretty much the size of Spain.

The sheep station had around 35,000 merinos, and the area was so huge they had to send up a spotter plane to find the animals (and occasionally new staff), and then we would have to go out and round them up using 125cc motorbikes. The holding paddock was more than 3,000 acres, about the same size as the whole of Knepp. When we got them up to the sheds, we'd hand them over to the shearers, dip them and then have to let them all go again. It took about six weeks in all.

Life in the bush was quite brutal in many ways, and I got a lot of stick from the hardened ranch hands as the 'new boy'. That was when the resilience I'd built up by going to thirteen different schools kicked in. There were a couple of fist fights when people tried to take advantage by giving me the dirtiest jobs, but once you showed you weren't a pushover it all settled down and you were accepted in the pecking order. I learned to slaughter sheep with a knife, and then skin them and do all the butchering too. Another job, which wasn't fun, was what they call 'mulesing', something they did to all the merino lambs at the time. It involves cutting off a strip of wrinkly skin and wool a few inches wide on each side of the sheep's anus. There's a lot of blood, but the idea is that scar tissue forms and the wool doesn't regrow in that area. The aim is to prevent the fleece under the tail gathering faeces and attracting blowflies that lay maggots in the skin, which will kill the animal slowly and painfully. Given the size of the sheep farms, the animals often go for long periods without being checked over. Many farms have now given up the practice, and selective breeding has produced merinos that don't have the same wrinkly skin around the rear end.

It was a rough sort of life, and there always seemed to be a lot of shooting – of emus and galahs (a kind of cockatoo), as well as culling of kangaroos and dingoes. We had to get a government licence to shoot kangaroos and then my job was to skin them and butcher them. The meat was injected with poison, a nasty chemical known as 1080 (its original catalogue number). Its chemical name was sodium fluoroacetate, invented in America in the 1940s but long banned in many countries and lethal to mammals if ingested. I'd be taken up in a small plane and have to sit in the back with all this smelly roo meat stuffed into sacks. We flew along at low altitude close to the cross-country railway tracks and my job was to drop the meat out of the open doorway. The idea was to kill the dingoes that were taking the farmer's sheep.

I also learned how to catch bulls on horseback and with a lasso while hanging on for dear life to a roll bar and perched precariously on the back of what the Aussies call a ute, a big land-cruiser-style jeep. I got to know how to castrate sheep and young cattle and dehorn them. The skill of roasting a monitor lizard in a fire pit was also acquired. I was taught how to crack a bullwhip, and how to find an emu nest. The eggs are fantastically beautiful, dark green like agate and as long as my hand. It was an incredible ten-month adventure, and hard work, but it was an insight into a different way of working on the land. There were good times, too: being flown to Exmouth on the coast with the boys and chartering a boat to go out fishing for giant trevally, and swimming for a bit of relaxation. Quite often I'd be sent out spotting in one of the little bubble helicopters and the pilot would put it down on a rocky escarpment where you could see for ever in all directions. The colours of the bush, and the rich red and orange rock formations were magical. You felt like you were the only person on some new planet, and that, quite possibly, no one had ever stood on that exact spot before.

I had my 21st birthday out in the bush, drinking billy tea brewed over the campfire and eating barbecued mutton. I don't know how long I might have stayed out there, but one day I got a message that my mother was very ill with what turned out to be cancer, and about 48 hours later I was back in England.

Once again I needed a job, and it was back to Knepp, where this time I was given responsibility for running the shoot. I knew the estate well and Charlie offered me a job as a gamekeeper. It was the beginning of a new life.

4
Poisoned Ground

I'm often asked if I have a favourite time of the year, or a favourite bird that I look forward to seeing when I'm out and about in the countryside. I never know what to say to that question, really, because I don't think of nature in that way. It's all connected in my head, and when I see a harrier standing high over the salt flats I'm watching how it's flying and whether it's being mobbed by other birds. I can admire its size and power, those wings stretching four feet across and the way the male and female birds lock talons and tumble through the sky when they are getting ready to mate, but I'm equally pleased to see a tiny little white throat darting into a hedgerow or a brown hare sitting in the grass. These are things I see practically every day here in Norfolk, but I don't tire of them. Each season brings its own rewards, and it's the movement and the signs of the cycle of the year that I find pleasing. Nature is about patterns and relationships, and making sense of the way everything is somehow connected. It's taken me over thirty years to try to understand those connections, and the way the seasons never fail to bring back recurring patterns that are somehow always subtly different. Nothing is ever quite the same as it was before. A change in temperature, more rain, a wind that blows unexpectedly from another direction – any of these things will have an effect upon the bird-nesting season, or the time the insects appear and what is around for them all to eat.

At Holkham, there is a large stand of very old pine trees, planted in the seventeenth century to try to keep the sand dunes at bay. From miles out at sea you can see them, a long, dark-green smudge standing proud of this flat land. Modern ecological science tells us that sand dunes don't need 'stabilising'; they move naturally and the forces behind them are too big for human beings to oppose. These pines – a mixture of species including Corsican, Scots, Monterey and Austrian – stretch for three miles from Burnham Overy to Wells-next-the-Sea. The trees were planted a century and a half after the first sea wall was created at Overy by controlling the sluice gates on the saltmarsh. By the middle of the nineteenth century, much of this formerly waterlogged land was being used for arable crops. The pines were planted to protect the farmland and grazing from the creeping sands. Even at the time, the estate's actions to change the landscape were debated, with local people resentful of the open vistas of sand and sea being obscured by a new forest. A disgruntled former employee of Lord Leicester, writing under the nom de plume of Dick Merryfellow, lampooned him in political pamphlets, dubbing him 'Prince Pinery'.

The pine forests have created their own distinctive

ecosystem, and provide a habitat for plants, insects and animals that don't occur elsewhere on the nature reserve, or necessarily anywhere else in Norfolk. Underneath these tall straggly species it is cool and shady, and pine needles carpet the ground. Pine forests are often thought of as fairly sterile habitats, the acidity of the pine needles stifling other growth and the darkness of the towering trunks inhibiting competition from other species. Over the years it has been suggested that the pines should be cut down and removed as non-native trees. But since 2019 a gradual programme of clearing of the scrub and some selective thinning along the southern edge of the wood has brought a growth in plant and insect diversity. Where there were once just a few foxgloves there are now clusters of bright flowers, their distinctive spikes of purple and pink trumpets standing tall on the forest floor. Orange tip and large skipper butterflies are common and the birdsong of blackcap and willow warblers cuts through the silence under the lofty trees. I have seen less common species too, like white-letter hairstreak, a delicate sandy-coloured butterfly with a distinctive 'lightning bolt' pattern a third of the way back from the edges of its wings. There is bright red campion in the undergrowth too, attracting hummingbird hawkmoths, those distinctive nectar feeders, stubby and chunky and seemingly unsuited to flight.

In some senses, the pine forest habitat is 'unnatural' and features species that wouldn't naturally have thrived here. But the distinction between 'non-native' and 'naturalised' is increasingly blurry. People tend to accept what has always been there, and more so when 'always' means what they saw as a child. Biologists are familiar with this concept – the 'creeping baseline'. We regard what we are accustomed to as the norm, even if it is an environment that is quite impoverished or altered compared to what our parents would have expected to see. Now, the pine forest has become part of what people regard as a permanent local fixture. Some conservationists believe that areas like this, containing species that would never naturally have grown in this environment, should be removed and 'nature' allowed to take its course. I would argue that much of what we have created in our farming landscape is equally unnatural. The real and complex challenge is working out what we can leave, what we can live with while still farming, and what should be altered or removed for the benefit of nature.

In early summer, a small, pale plant emerges under the pines. It's virtually colourless because it contains no chlorophyll, and visitors (those few who notice it) sometimes mistake it for a mushroom. Known as the yellow bird's nest or the Dutchman's pipe (*Hypopitys monotropa*), it gets its nutrients from the mycorrhizal fungi that live in harmony with the roots of the pines. We have found that careful management of the pine woods is vital to the survival of this unusual species, and that where the trees are felled to benefit other flowering plants they do not thrive. It's therefore important that we maintain areas of thick tree cover, with little direct sunlight penetration. This past year has seen a record number of the bird's nest fungi emerging from the forest floor where it is covered thickly with moss and pine needles. At about the same time, and in similar locations, we see the tiny hairy orchids known as creeping lady's-tresses (*Goodyera repens*). They also emerge in very restricted areas under the pines, but only where the trees are at least a century old. Just a few inches high and with flowers smaller than my little fingernail, they also live in symbiosis with the mycorrhizae that colonise the ancient tree roots. The mycorrhizae are essential to how many plants absorb nutrition, and they also play a role in the overall health of our soils. The mycorrhizal networks store carbon underground, and, even when dead, provide structure to the soil. While living, they work in symbiosis with the plant, providing nutrients like sugars to the fungus while in turn helping the host plant absorb water and other key nutrients such as phosphorus.

Neither the creeping lady's-tresses nor the yellow bird's nest are flamboyant, but they are interesting precisely because they are so dependent on the specific environmental conditions here in the pine forest, and because they lead such secretive lives. Holkham surveys and counts these populations, and maps their precise locations so that we can try to preserve the best conditions for their survival.

Not far away, where the dunes lead out towards the beach, there is another cryptic creature – the antlion (*Euroleon nostras*), an insect which, in Britain, is only found in Suffolk and Norfolk. Just about an inch long, they are almost impossible to see on the wing as they fly at night, and look just like a cross between an ant and a dragonfly, with a double-pair of long, elegant lacy wings. When they are at their larval stage they resemble an armoured tic and they hide just under the surface of the sand, where they dig a conical depression on the surface. It creates a cup shape that looks just as if someone has pushed a tennis ball into the sand. When an ant, beetle or any other small invertebrate falls into the depression it tries in vain to climb out, but the desperate flailing of its legs just causes the sand to avalanche back into the pit. The submerged antlion larva detects the vibrations of its victim as it tries to scramble up the sides of the bowl, then pokes its head out from its hiding place and flicks sand at the victim, knocking it back towards the bottom of the pit. Then, a pair of tiny pincers emerges from under the sand grains and grasps the prey. It uses the prongs to inject the victim with venom to liquefy its organs so that they can be sucked out. Afterwards, it flicks the withered corpse of its victim out of the sand trap so as not to betray its presence. In July and August these tiny life and death struggles play out mostly unseen along the dunes.

One of the other questions I'm frequently asked is what makes me care so deeply about making the countryside a better-functioning ecosystem. I don't have an easy answer to give. I suppose that growing up in and around nature, and being encouraged to be outside by my mother, who probably recognised that I was happier there, is part of it. There was never a long-term plan.

After Knepp, when I had gained some experience, I took a job as a second gamekeeper at an estate on the Welsh–Shropshire borders. It rained a lot, and the shoot that I was helping to manage was very commercial and successful. For personal reasons I wasn't especially happy there, and within little more than a year was keen to move. That was how I first came to Norfolk and a job at Raveningham, which has been the home of Sir Nicholas Bacon and his family since 1735. In a way it was a sort of homecoming to East Anglia, mirroring my parents' early married life in Suffolk. I don't feel I come from anywhere in particular, but if there is a place I feel at home, then it is here, and both of my own children were born here. I like the relatively unpopulated landscape and its sheer variety, with everything from the sea and marshes to woodland and chalk streams, and although it's famously flat there is in fact a lot of rolling topography.

'Rav', as everyone there calls it, is much smaller than Holkham, probably a quarter of the size at a little over five thousand acres. The farmland is mostly used for arable production: wheat and barley, sugar beet and potatoes. There are cattle and some sheep, and a lot of ancient natural woodland covering around 400 acres with stands of ash and oak, hazel and sycamore providing a lot of diversity and good habitat for orchids and butterflies like the white admiral and pearl-bordered fritillary. Wild game shooting was always of a very high quality, but just like everywhere else in the 1980s and

1990s, they had started to see serious declines in the number of birds that successfully fledged. The soil is different there to Holkham, with more heavy clays, especially towards the south.

I grew to know Rav like the back of my hand. I started as second keeper there in 1995, but a few years later rose to be head keeper and ended up staying there for twenty-three years. My job was not supposed to be about conservation, certainly not at the beginning, but to make sure the estate delivered a healthy surplus of wild game. As a gamekeeper, you need to be out on foot, watching, always watching – what's happening around you. You watch the weather, you watch what's growing and what's moving. Movement is what triggers your eye. You learn to spot the smallest flickering things in the landscape. You don't recognise a bird silhouetted against the sun by its colour, but by the shape of its wings or the way it rises from the ground if you disturb it. And then you start questioning things: most of the time an animal isn't moving at random, any more than a plant is growing in a particular spot for no good reason. Why is that crow sitting on that particular fencepost and not the one next to it? The way things move tells you what they are, even when you are too far away to make out colours or patterns. My son Nathaniel has memories of coming out with me at dusk at Raveningham and hiding in ditches to watch for barn owls. There was one spot, in a field called Low Barn, where I would use an old gamekeeper's trick of making a sound like a rabbit in distress. You do it by making a 'trumpet' out of your closed fist and blowing out through pursed lips to make a disturbing high-pitched squeal. The owl would sometimes come right over our heads, a ghostly apparition drawn in the fading light by trickery. Once we even managed to call a stoat which was drawn to investigate the noise and it came so close it practically sniffed our boots.

I recall clearly that it was a warm autumn evening as I walked out over one of the estate's larger fields, covering somewhere around twenty acres. This was the magic hour, the time in the evening when nature becomes active and things come out to feed before it gets dark. There are two of these so-called magic hours in the day, one in the early morning and one just before sunset. In early September, the evening magic hour is around 7 p.m. in Norfolk. This particular field was flanked on the southern edge by an ancient, semi-natural patch of woodland known as the North Belt. The stand of trees contained wonderful oaks two, three and four hundred years old. In spring it turned into an enchanting bluebell forest and there would be carpets of strongly scented, rather noisome dog's mercury. On the western edge there was another wooded area, much more recently planted and dominated by sycamores. The air had that distinctive autumn flavour, a slightly cool edge that told you the summer was really over and the heat of the sun would not be coming back. From here I had a view of fields gently rolling away to the east. It was a favourite spot with somehow a different feeling to other parts of the estate.

I remember there were skylarks singing sixty feet

above my head and the hedgerows were bearing fruit. There were still some late swallows and house martins feeding, fattening up their second broods ready for the long flight south. I could hear wood pigeons pootling about in the stubble. A hundred yards or so to my left I noticed movement at ground level, something unusual. As I walked through the stubble I realised that it was a brown hare, but it wasn't moving away. It was lying on its side, its limbs making all the actions of running, and the claws on its feet were scratching an arched pattern in the soil, like children do when they make a snow angel. It didn't seem to be injured, but there was something clearly wrong. This animal was writhing and in pain, its gummy eyes somehow blinded. I stretched its neck to kill it humanely. And then, I saw other moving things elsewhere in the field. Each one of them was another hare, and each time I acted quickly to end their suffering. I had never seen anything like it, and I suddenly felt a crushing sense of despair. As I killed each successive animal I felt like crying. That doesn't happen easily to me, but there was something terribly wasteful and destructive about whatever had happened to these animals. For so many people the hare represents wisdom, something eternal and deeply powerful about the countryside. Going back to ancient times in Britain they have been revered, imbued with changeling powers, and seeing them is a simple joy. I wanted to get to the bottom of what was killing them.

The next day I went into the farm chemical store and started looking at the containers which had held the herbicides that I suspected had been used on the fields. I read the labels on the large barrels containing the sprays that promised to desiccate the remaining vegetation after harvesting. It's a process farmers have come to rely on to 'clean' the field and make sure whatever is planted next doesn't get contaminated with fresh growth from whatever was there before. Desiccants are herbicides which kill only the parts of the plant they touch, making them die back or dry down rapidly, often within a day or two. The culprit, I soon realised, was a chemical called paraquat (finally banned in the UK in 2008). It's lethal to human beings in tiny doses and the hares had ingested large doses, accidentally of course, by eating the freshly sprayed green leaves.

It wasn't my job to go and read the labels on the farm's chemical sprays, but I often wonder why so few people do make the effort to look around them and make connections with what they see in the natural world. One of the things that's happened with the increased mechanisation and technological advancement of farming is that fewer people are actually out there walking the fields and really seeing, smelling and touching their environment. Machine operators are often sitting in a very comfortable cab, sealed off from the substances they are spraying and listening to music either on the cab sound system or through their wireless ear pods. They have a clinical view of the fields around them, protected from the elements, the noise of the tractor and of the chemicals they are dispensing on to the land.

I've been asked why I should find the deaths of a few

hares so upsetting. As a gamekeeper, my career was, in one sense, all about killing. I've trapped stoats and shot foxes as well as hundreds of rabbits, partridge, duck, grouse and deer. I've hunted wild boar in France and stags in Scotland. But that was always for a purpose. I didn't get pleasure from shooting a fox, or any other animal or bird, but I did get a feeling of satisfaction. Why? The birds were always eaten, and in the case of a fox, I knew that removing a particular individual from a marsh would very quickly benefit breeding lapwings and their eggs. Last year, on one of the lapwing fields at Holkham, we had an incursion by a vixen and within 48 hours we had lost 70 per cent of the nests. A fox will return again and again once it finds an easy food source, and this one had found the nests at the right time when there were eggs available. Many species do this; I've noticed a heron near my office that specialises in catching moles. Now that it knows how to catch them it will carry on doing it. Finding the fox is like a game of chess; tracking it and shooting it cleanly with a rifle requires fieldcraft and skill. If I shoot it from three hundred yards away it doesn't even know I am there, or that it is being targeted. There is no distress or suffering, no exhausting chase or fear. But on that lovely field at Raveningham those brown hares were not being killed deliberately. They were lying there, dying in agony, and were simply the unintentional casualties of careless, economically driven chemical farming. Seeing their wanton destruction, and their agonising deaths made me question why we were happy, it seemed, to spray so many dangerous poisons to the crops that we would later be eating.

Chemicals have their uses, but there has been a degree of acceptance about their widespread application that I find worrying. It's a huge industry, dominated by a few key players, namely Bayer, BASF, Syngenta / ChemChina and Corteva, which produce three-quarters of the world's pesticides and more than half of the world's agricultural seeds. Further investigation reveals that the chemical companies are often part-owned by a small number of (almost exclusively US-based) private equity firms, which in turn are major investors in the relatively small number of giant food producers. They promise much, even though a recent 2021 report from the Centre for Biological Diversity and others showed that the EU was subsidising pesticide production with larger amounts of money than the sector was making in profit. The study shows that pesticide growth and the drive for an increasingly intensive agricultural production system have gone hand in hand, supported by governments spurred on by a nebulous understanding of food security. Within this model, there are what the study calls four main pillars: industrialised agricultural machinery, synthetic fertilisers, hybrid seeds and widespread use of synthetic pesticides. These pillars have contributed to increased agricultural yields, but at what cost? Yes, average global yields have increased by 50 per cent in about thirty years, and many agricultural diseases have been kept at bay, but crops – including maize, rice, wheat and soybean – are now seeing declining yields in spite of increased chemical applications. The report attributes this largely to pesticide resistance, but also to climate change, soil degradation and biodiversity loss. We are now starting to understand the widespread environmental impact of that chemicalised system, even if we can't produce a cash balance sheet to assess its financial 'worth'.

In this country, Defra research shows that many farmers take a risk-averse approach when it comes to what chemicals can do. Rather than wait to see if they have a particular pest or disease, they spray anyway. And again and again, subsequent studies have told us that there are negative consequences for the natural world in using these chemical weapons. The downsides are often not immediately apparent, and only really reveal themselves as these seemingly helpful innovations become widespread. A lot of the damage caused by supposedly progressive inventions and techniques in agriculture take years to appear. It was only in 1962, when the American biologist Rachel Carson published her groundbreaking book *Silent Spring*, that large parts of the world finally woke up to the dangers of DDT. She had been trying to alert the world to its dangers for many years, but magazines had rejected her articles on the basis that they were 'unpleasant'. Carson's book and articles in the *New Yorker* were famously read by President Kennedy, who took her warnings seriously enough to order a scientific enquiry. Arguably, her work was one of the key spurs for the modern conservationist movement. *Silent Spring* sparked a massive public debate – Carson was accused by one prominent contemporary biochemist of being fanatical in her defence of what he called 'the cult of the balance of nature' – but also an awareness of what was being done to the natural world in the name of progress. However, although there was a voluntary ban which many farmers observed much earlier, DDT was not banned for agricultural use in Britain until 1984.

Paraquat is no longer sold in the UK, but there are other chemicals being applied to the soil and the vegetation which are widely implicated in the loss of our insects and birds. Much more recently, the debate over the use of neonicotinoids and their potential contribution to the decline of bee numbers raged for several years. It may not be possible to farm absolutely everything that we need organically, and there are serious cost implications to trying to do so, but quite often a bit of planning and forethought makes it possible to grow crops without always using chemical treatments automatically. Often we spray chemicals prophylactically, and that's disastrous because the organisms they were designed to combat start developing resistance. Soon, we're back to square one, looking for a new, more powerful chemical to do the same job.

Recent figures from Defra show that between 2011 and 2019 there was a 7 per cent increase in the area of land treated with pesticides, although there has been a slight drop in the last couple of years. They include soil sterilisers, fungicides, insecticides, molluscicides, repellents, sulphur, growth regulators, biological control agents and physical control agents. The list of active chemical ingredients occurring in these branded products is much longer, and it's very hard for farmers to know how best to use them. They rely on advice from government, and they assume that the licencing authorities are making the right decisions. One of the things I saw under the CAP system was the madness of farmers being ordered to 'set aside' ten per cent of their land to discourage overproduction. But at the same time, the directive came through that allowed them to spray and mow their fallow fields in the spring, meaning, everything that might be using that 'unfarmed' land to nest on would just be killed in the spring. Who dreamed that up?

In the mid-1990s we were using pesticides at an alarming rate. In his meticulously researched book *The Killing of the Countryside*, Graham Harvey calculated that at that time every acre of winter wheat in Britain was being dosed with eight treatments of pesticide, including three applications of fungicide. Harvey's book struck a chord with me when it first came out in 1997, and it coincided with my own realisation that a lot of the intensely destructive things I was seeing in the countryside were happening all over England. For potatoes, to take just one example, the number of applied treatments could be as high as twelve. At this time a massive quantity of sulphuric acid – around 33,000 tonnes per year – was being used as a desiccant for potato plants. Since then, according to Defra, measured by weight, the total quantity of pesticides used in British agriculture has gone down, but the number of applications of more potent chemicals has increased. The EU is also calling for pesticides to be reduced by 50 per cent by 2030 as part of its 'Farm to Fork' strategy – its own carbon reduction plans (net zero by 2050) as outlined under the European Green Deal.

The British government has said it wants to put non-chemical pest control at the heart of its 25-year Environment Plan, acknowledging, among other things, that farming is the single most significant source of water pollution and ammonia emissions (especially from slurry and manure) in the country. Farming contributes 25 per cent of the phosphate, 50 per cent of the nitrate and 75 per cent of sediment loading in the water environment, all of which we know harms the ecosystem of rivers and lakes. Phosphate – in the form of extracted rock minerals or made from animal bones and waste – is a major cause of pollution in rivers and lakes, causing eutrophication (unnatural enrichment), which results in reduced oxygen levels and the excess growth of harmful algae. Although phosphate fertiliser use has declined in recent years, farmed soil carries a significant historic load, which is still being released as rainwater drains off the land. Nitrate pollution also reduces oxygen levels in water and affects the quality of our groundwater, with agriculture a key source due to artificial fertilisers containing urea, ammonium and nitrate. The UK does not have a good record when it comes to water quality, especially in southern England. Defra concludes that agriculture is a marginally bigger contributor to inland water pollution than sewage and waste water. In their studies, only 14 per cent of our rivers are classified as 'close to their natural state', despite a 60 per cent reduction in the quantity of phosphates and 70 per cent fewer phosphates entering our rivers from waste-water processing over the past thirty years. Worryingly, while 21 per cent of English rivers were accorded 'good' ecological status in 2021, none were classified as having good chemical status, and their reports refer to rivers containing a 'chemical cocktail'.

There is promise of an integrated pest management scheme to encourage using chemicals more judiciously and supplementing them with improved crop husbandry and the use of natural predators. And yet despite this, the trend in recent years for chemical use has been upward. It's estimated that the land used for crop growing in the UK is around 4.5 million hectares (just over 11 million acres), but the quantity of land being sprayed in total rose from just under 60 million hectares in 2000 to over 73 million hectares (180 million acres) in 2016. These larger figures are called 'spray hectares' and account for the fact that each field will be sprayed numerous times in one growing season.

More worryingly, the number of different chemical products used rose from twelve to sixteen in the same period. And remember, one proprietary chemical product will often have numerous individual ingredients. The most commonly applied fungicide in the UK, chlorothalonil, used principally to prevent mould and mildew on barley and wheat, was banned across the EU and in the UK in 2020. It has been widely used across Europe and the USA since 1964, which, ironically enough, was not long after *Silent Spring* was published. Regulators found it caused serious problems for fish and amphibians and was implicated in the decline in bumblebee populations. There were also, according to the European Food Safety Agency, worries over its possible effect on human health and its potential to cause changes in DNA.

I saw all of this happening in Norfolk at local level. At Raveningham, there was a farm manager who was very keen on productivity, and under his leadership the estate shifted away from growing cereals to crops like carrots, parsnip, potatoes and fruit – all of which were more lucrative, but which also needed more intensive types of production. I saw hedges being flailed, fields sown right up to the edges and doused heavily with chemicals. There was very little space left for nature.

As gamekeeper, my primary concern was ensuring that there was a surplus of birds for the shoot, principally pheasants and English partridge. Just like most other estates, Raveningham had seen years in the 1990s when the 'game bag' – the tally from a shooting day – had severely declined, and on some occasions ended up with no birds at all. English, or grey partridge (*Perdix perdix*) were already down to one breeding pair on the estate when I arrived. These charming little birds have large clutches of eggs – easily fifteen or even more. Once hatched, partridge pairs will collaborate and band together with other pairs and their young to form a covey; these little family groupings will stay together through the autumn and only break up when the adults start pairing up again in January or February the following year. A hen partridge usually only lives for eighteen months – that's two breeding seasons – and within that short period, they need to maximise their reproductive potential. Sometimes a pair of birds that have lost their own chicks will join up with another hen and travel together as one covey, sharing the 'childcare'. This year, I have been watching a covey of twenty-five birds on Great Farm. They are nervous creatures, but the site of the tiny chicks following the hen along the edge of the crop never fails to raise a smile. The chicks grow quickly and can fly after only about ten or twelve days, but they are very dependent on developing their feathers quickly once hatched. That only happens with the correct diet.

The BTO recorded a more than 90 per cent decline in partridge populations from 1967 to 2010. Whereas many estates had been artificially rearing birds for some years, giving partridge eggs to chickens to incubate and then look after for the first few weeks of their lives, at Raveningham we had not done that. In my view it was a poor strategy, as the partridges didn't learn the right skills from their 'chicken mothers' to survive in the wild. Alongside the greys, we also had red-legged, or French partridges (*Alectoris rufa)* which had been introduced to England in the late eighteenth century from the Mediterranean. They have been less affected by the decline in cereal insect numbers than the native greys, because they seem to eat more seeds than vegetation. The red-legs are slightly larger, and more flamboyant in their plumage.

They have different personalities, and they are much worse parents than the greys. They wander around the fields with their chicks following them haphazardly, and are much less protective. They don't collaborate with other adult birds like the greys do, and tend to lose many more of their young. They may have been here for 250 years, but they haven't established a resident population, and are almost all bred artificially. For some years, they were also crossed with the Asian chukar species (*Alectoris chukar*), resulting in a hybrid bird which in turn has raised fears about the long-term survival of the wild birds. From a hunting perspective, the greys are valued more highly as sport. They are much harder to shoot, because they come over a hedge as a covey and spot the guns. Then the group separates in a wild starburst going in every direction. Also, in my view, the reds don't taste so good, but they were bred even more intensively in the late 1960s as a way of supplementing the decline in English partridge and other game bird numbers brought about by intensive farming and chemicals.

Regardless of how they are reared, grey partridges have very specific feeding habits: they need to eat around 2,000 insects a day in the first few weeks of life and thrive on fat, softer-bodied species like aphids, weevils and sawfly larvae. They like to forage on semi-open ground within easy reach of crop cover with a good canopy to hide them from predators. That cover mustn't be too dense at its base so that they can run through it easily. Denser crops also hold rainwater for longer, and a key threat to very young birds is getting chilled through being too wet. It's pretty clear that insecticides and pesticides have also been hugely damaging to the survival of grey partridges. Crops which have been sprayed to keep weeds at bay will harbour many fewer insects, and will also grow more densely.

To try to help the game bird population at Raveningham, there were several elements to my strategy. The intensive farming had left us with no winter stubble, and the planting schemes were monoculture followed by a root crop then another monoculture. On the plus side, Sir Nicholas Bacon, and his father before him, had resisted attempts to maximise the size of their fields. They had largely held on to hedgerows and also to ditches and dykes which are – in effect – hedges in reverse, providing dense growth, harbour and acting as a reservoir for insects. There was one area in particular named the 'Onion Field', which was actually three fields separated by ditches. As part of the intensification of the farming cycle, the ditches had been filled in to make the land easier and more efficient to harvest. Meanwhile, in other places, the intense production cycle had led to a decline in the health of the existing ditches on the marshes. Excess nutrients from fertilisers running off the crops had caused thick weeds to grow out of control and fill up the dykes. They had become so clogged that we began to fear that if cattle fell into them, they would get stuck and drown.

It takes about six years to return some kind of natural balance to land that has been farmed intensively, what I usually call 'Taliban-style farming', which kills everything it doesn't want. One of the things I saw then, and still see in other places, is the application of pesticides according to the calendar, rather than according to need. When I was at Raveningham, they were spraying the wheat with what's called 'ear wash', a fungicide designed to protect against insects like orange wheat blossom midge or fungal diseases like fusarium and microdochium, as well as other moulds and brown rust. These sprays are applied to the emerging ears of corn and wheat just when the young fledgling birds need their peak intake of insects. I would always walk out among the crops, and it was clear to me that there were dead insects all over the ground when the sprayer had been in action. I would get down on my hands and knees and find the corpses of flies and spiders, wasps and all manner of smaller things. What I hated was the idea that you sprayed no matter whether you had evidence of disease or infestation, because that's what the chemical companies wanted you to do – in order to sell more of their products.

Once I had the go-ahead to take some of the Raveningham land out of food production – essentially the fields that were uneconomical – we saw remarkable changes. In all, it added up to about a thousand acres, or 20 per cent of the farmland. Very quickly, plants that had been absent, like marsh foxtail (*Alopecurus geniculatus*), started flourishing, and we noticed that after flowering, when the seeds became available, they attracted thousands of widgeon to the marsh. On the land set aside for nature we used no fertiliser, no fungicides and no pesticides. The absence and shortage of wild games birds was reversed. After a few years we had hundreds of grey partridge, a decent shootable surplus, and of cock pheasants too. The pheasant shoot had been barely viable when I arrived on the estate. When guests came to shoot pheasants, I was in the habit of telling them that their priority was to shoot cock pheasants, not just to kill things for fun. After watching how the game bird population increased, it was very pleasing when HRH the Duke of Edinburgh presented Sir Nicholas Bacon with first prize in the 2004 Purdey Awards. The prize recognises those who have promoted a wider understanding of game and habitat conservation work.

As the years went by, I would go to neighbouring farms to talk to people about what we were trying to do. It was very satisfying to speak to one 70-year-old farmer who took a field out of production and to hear him say the following summer that he had seen brimstone butterflies – a species he hadn't seen since he was a boy.

In 2002, I persuaded Sir Nicholas Bacon to let me bring in the digging machines to a 140-acre patch of land that had been deep drained in the 1970s. We turned it back into a wetland, but we used it as somewhere we could also graze cattle. Something incredible happened: snipe, marsh harriers and lapwing started turning up, and then nesting. The BTO had calculated that a healthy reproductive rate for lapwings was 0.7 fledged chicks per nest. In the second summer after restoring the wetland our lapwings were achieving 1.4 chicks per nest, exactly double. In addition, we saw an upsurge in Redshank, and noticed that young birds that had been born and ringed about ten miles away on the RSPB reserve at Breydon were settling on our wetland.

It's crucial to remember that Raveningham remained a profitable farming business alongside these environmental changes. What we achieved with habitat restoration and regenerative farming didn't result in the estate losing money. It was the opposite: the yields improved, and revenues also rose. Restorative agriculture isn't about setting aside a chunk of land and just letting it go back to nature – that doesn't work. We have to manage it properly to provide the habitats that work best with the species we want to help. The principles of regenerative farming are not complicated. But they do involve a change of mindset, and the understanding that our soils do not need to be impoverished as we farm.

The idea of not disturbing the soil is one of the founding principles of regenerative agriculture. We are still learning about the complex network of fungi that live beneath the surface, and how ploughing disrupts this living system and also releases carbon into the atmosphere. Ploughing also disturbs the structure of the soil, often removing the natural aeration created by worm burrows. Using 'cover crops', especially over the winter, is another key principle, choosing species like clovers and vetches that capture nitrogen and replenish the soil. Gabe Brown is an American farmer who started changing the way his land was used, having watched his profits diminish every year while following the standard model of adding fertilisers and ploughing his fields. He calculates that there are 32 tonnes of free nitrogen above every acre of land – and we should use it. In his book *Dirt to Soil* he says that few people, including farmers, have truly respected the notion that the soil itself is its own ecosystem, one which functions in extremely complex ways. He realised, among other things, that when he stopped ploughing and adding fertilisers to his soil, it became more fertile over time. Brown postulated that the addition of synthetic fertilisers was disturbing the relationship between plant/crop roots and the mycorrhizal fungi that acquire minerals for the plants in exchange for carbon. Synthetic nutrients tend to be of only one type, not the full range that the healthy plants need.

One of Brown's fundamental rules is that the disturbance of soil should be minimised in order to preserve its healthy structure. When soil is ploughed or turned, oxygen is infused into the soil, which stimulates the growth of bacteria which in turn consume the soluble carbon-based 'glues' that hold the tiny aggregates of sand, silt and clay together. This then causes the soil to become less porous, anaerobic and with less nitrogen. Carbon is also released into the atmosphere. Brown says that as plants put down roots, they take in carbon dioxide from the air and combine it with water to make simple sugars, known as photosynthates, which in turn promote growth. The plants' roots also exude chemicals into the soil as nourishment for microbes. One of Brown's most astonishing claims is that some plants use 96 per cent of the carbon they process to feed soil fungi and bacteria. How this all works together to enrich soil is still poorly understood in many ways, but the topsoil, where most of the rapid biological activity takes place, is very sensitive to disturbance – whether that be tilling or the addition of chemicals. We are learning that plants, somehow, have a natural mechanism which sends out signals within the active topsoil, seeking out the correct nutrients and attracting them. It seems sensible to encourage this by not leaving soil denuded even after we have harvested our food crop. Bare soil, says Brown, is rarely healthy soil. Cover crops also protect the soil from excess heat and cold, and prevent rainwater washing away the soil.

Creating diversity is also important as it supports a wide range of organisms, and a diverse ecosystem is more resilient. If land is only used to grow one species, then anything that threatens the health or survival of that species has the potential to kill it all. Brown also reveals that the diversity, even of farmed crops, is much, much less rich than it used to be. In the USA, for example, in 1900 there were more than five hundred varieties of cabbage available to consumers, whereas today there are fewer than thirty. That pattern has been repeated over and over again as agriculture globalises. The danger with specialising and reducing variety is that it impoverishes the biodiversity available for other species, some of which may depend on plants that we no longer consider commercial or useful. All too often we put all of our eggs in the same basket, leaving us vulnerable to a natural catastrophe, some disease that targets one species more than others. Once a species is driven to extinction we have no other options – and whatever 'ecosystem services' it provided are gone for ever.

Keeping some grazing animals on the land is another tool for the regenerative farmer: using fields for pasture instead of always planting a new harvest crop gives the soil a chance to rest, and the animals provide fertiliser. The process of being grazed in turn keeps the plants stunted, and from using the carbon stored in the roots for further growth and seed production, making them easier to control.

For me, wildlife is just like the other products on the farm: it needs to be 'grown' efficiently and with direction. It's been shown, by me and by others, that making space for nature doesn't necessarily lead to a decline in food production. Nature-friendly farming actually reduces the negative impact of intensive agriculture by creating better biodiversity. Many of the things that live or grow in nature-rich areas actively enhance food production by increasing the number of insect pollinators or, in some cases, by fighting off predators which could harm the crop yields. A six-year study on the Hillesden Estate in Buckinghamshire (which coincidentally once belonged to Coke of Norfolk), showed that taking up to 8 per cent of land out of crop production caused no net loss of yield. Some crop-density reduction occurred at the field edges, but overall the yield of the farm went up. That study, the results of which were published by the Royal Society, was completed a decade ago, and the scientists behind it made the point that better engagement with farmers and indeed training them in how to do the same, are essential if we want to improve the quality of the natural habitats on our agricultural land.

Sometimes it just takes a very small tweak of the environment to bring spectacular results. At Raveningham, when my children, Teale and Nathaniel, were young, we were living in a gamekeeper's cottage and I began to experiment with planting some hay meadow wildflower species along the hedge that divided our garden from the field. The following year I started noticing a few orchids creeping into our strip from the nearby woods. My daughter, Teale, who was about ten at the time, started playing a game of counting the cowslips before supper. In the first year we counted eleven; the next year it was twenty-seven. By the third year it was thirty-five and in the fourth year there were too many to count.

May 2020
Great Farm

It's the first sunny day after what seems like weeks of cloudy, dark and dismal weather. This year has been odd: five weeks of drought followed by the same again of heavy rain. Great Farm has a new pair of gates with shiny hinges, and as they swing open I smell the unmistakeable scent of fresh creosote. It's a complex combination of something toxic, very chemical and yet evocative of the farmed environment – timber buildings and gates that need to do a serious job out in the elements. The warmth of the sun is bringing it out, cooking it on the wood and making it sticky to the touch.

I'm driving on the new track, and as I crest the slight incline and head down alongside the new field margins I have a sense of beginning, of hope that comes with the quickening season. To my right there's a small copse that needs thinning. It hasn't been touched since it was planted in the 1960s and the softwoods, mainly conifers, have strangled the beech and oak saplings. There's some open ground at the lower end and I can see that there's a single oystercatcher nesting there. But it's the hedges that strike me. In February they were bare, and somehow dark and wizened under a coating of early morning frost. They're not brutal flat-tops any more, and they have colour from the hawthorn blossom that's been allowed to come through. I can see thistles close to the hedge line, a good sign as they will provide pollen for many pollinators, as well as winter food for the goldfinches.

The hay meadows were drilled ten days or so ago. I'd like to see if there's any sign of the new flower species that will help change the biodiversity of this farm. The first thing I notice is that there are still some recruit potatoes, remnants of the crop from three years ago, that are still emerging. I'm relaxed about it because the meadow areas will be mown short soon, part of the process that gives it the best chance to flower next spring.

To my right now there is a cultivated area – the soil loosened and broken up to encourage what farmers call 'annual weeds'. They're the species that stood high in the meadows when I was a child, a long list of plants that attract pollinators and insects. There will be cornflowers, poppies, corn marigolds, field pansies and Venus's looking-glass. They're in the seed bank and, given time, will emerge and express themselves. From a distance, and thanks to the steady recent rain, I can already see a green haze above the cultivated band around the field. It's a precursor to what I know will be a feast of natural colour, and it will change the way Great Farm feels.

5
The Business of Beauty

Sometimes, making space for nature is about making small tweaks to things that don't cause us any real inconvenience. One of the aims of regenerative agriculture is to make parts of the countryside more accessible to the general public. And one of the lessons of the coronavirus pandemic is just how desperately people need to be able to walk and get out and about in nature. For many landowners, the Covid-19 pandemic was the first time they had had to cope with very large numbers of walkers using footpaths, and in some cases straying from permitted rights of way. Erosion and damage were significant, and a lot of farmers are frustrated by the amount of litter and dog waste that was left behind on their land. As our population grows, these issues are clearly going to become more significant.

At Holkham, the majority of the visitors want to see the beach. It's consistently voted one of the best beaches in the whole of the UK, and it's a persistent draw all year round thanks to its seemingly endless stretch of sand. Summer and winter, the great canopy of sky is mirrored by the moody reflections thrown off the North Sea, and the dunes are a tranquil backdrop to the drama of the sweeping emptiness of the beach. I think people are drawn to it by its endlessly changeable light, and the sense that the horizon seems limitless. The brisk winds coming in off the North Sea seem to have the power to cleanse the mind. In winter, when there are many fewer human visitors to the sands, there are sometimes little groups of snow buntings busily darting about at the fringes of the dunes. The tiny arctic birds are hard to spot on the ground, well camouflaged in their less-dramatic winter plumage on the shingle or against the grass until they take flight, the males showing the dramatic bright white wing patches that mark them out from other finch-like birds. As Christmas approaches and the cold sets in, massive armies of gulls may arrive to feast on hundreds of thousands of razor clams that can appear on shore without warning.

It seems clear that the pressures on our relatively empty areas of countryside are going to increase with time. Our population may be increasingly urbanised, but in peak holiday periods that may lead to a proliferation of visitors at nature reserves, especially where they offer a winning combination of attractions, such as glimpses of the natural world coupled with activities of one sort or another. Although the scale and intensity of all this may be something that is very hard to manage, it's important to realise that it's actually not a new problem. In 1945, the House of Commons received a report on the creation of National Parks by John Dower, Secretary of the Standing Committee on National Parks, who said:

'There will have to be, from place to place, some sacrifice of those scenic delicacies which are only possible "among the untrodden ways".'

One of the challenges of managing the Holkham Nature Reserve, much of which is actively farmed, is coping with the 800,000 people who visit the beach each year. They arrive in 500,000 cars and bring with them about 300,000 dogs. That's a lot of dog-poo bags to deal with, and in summer we expect to have to dispose of one tonne of dog waste each week. It ends up in landfill – no one wants it – and it can be quite toxic.

The other problem with dogs is that they can very easily disturb, and in many cases kill, ground-nesting birds. At Holkham, the sand dunes and certain areas of the beach are key sites for nesting oystercatchers, little terns and ringed plovers. All three species lay their eggs on the ground in shallow scrapes, sometimes on shingle or gravel, or in the dunes. They rely almost entirely on camouflage to protect the eggs, although they will valiantly try to defend their clutch against even human-sized intruders who get too close.

It's important to note that while we try very hard to minimise our human impact on these populations, it doesn't mean we can artificially protect them from natural hazards. Last year, at one stage in spring we had more than thirty little terns nesting behind the cordon at Holkham Bay. In little more than 72 hours, 90 per cent of the eggs were eaten, not by dogs, but by an otter. However, we always need to remember that on top of the natural pressures within the ecosystem, larger and larger numbers of people in these environments puts yet more strain and stress on these wild creatures. We can't eradicate otters and foxes, or even predatory bird species – and we wouldn't want to, unless there was an imbalance. But when we add human pressures to those wild hazards it can often tip the scales into total catastrophe.

I'm a dog lover, and I don't want to stop anyone exercising their dog. I decided that the way to minimise conflict between the nesting birds on the beach and in the dunes was to try to gently persuade people to change the way they behaved. It's always been a rule that people in the countryside – or anywhere in public – should have their dog under control. That doesn't mean that it must always be on a lead, but it does mean that a free-running dog should obey commands and reliably come back when it's called. As we all know, that is often not the case. We opened a period of public consultation and it led to a seasonal zoning programme for the beach. The result was an area of 120 acres, just 20 per cent of the beach, being put off limits to dogs, but they were not banned. More than 70 per cent of the beach at Holkham remained open access. We erected a series of rope cordons with signs asking people to stay out of those zones and placed signs well ahead on the approaches warning them that restrictions were forthcoming. The controls were in place from 1 April to 31 August. The results were impressive, and the beach wardens found that only tiny numbers of people ignored the signs, and when a dog did penetrate the cordons it was usually by accident.

Some conservation organisations have imposed blanket bans on dogs in their nature reserves, and it's usually a bone of contention, with feelings getting very heated on both sides.

I found it interesting that when we held the public consultation on dogs at Holkham, there were more than two thousand responses from the general public. That may sound like a small number when we have 800,000 annual visitors, but I recall that when I sat on the panel for the National Parks Designated Landscapes Review, we received only marginally more, and that covered every single English National Park and Area of Outstanding Natural Beauty. That review was completed in 2019, under the chairmanship of Julian Glover, alongside Defra officials and the other members of the panel: Ewen Cameron, Jim Dixon, Sarah Mukherjee and Fiona Reynolds. It was an attempt to try to better understand and shape how we can protect and enhance the roughly 24 per cent of England that is covered by 'national landscapes'. The review found that there was a lack of coordination in the way National Parks and Areas of Outstanding National Beauty operate, and that they were 'not delivering on their duty in relation to nature'. The review was clear in its recommendations to the Secretary of State that these types of designated landscapes could (and should) become leaders in nature recovery. The panel were very willing to listen to my suggestions about agriculture, and accepted wholeheartedly the notion that farmers need to be treated as partners in these relationships.

Even before Covid-19, it was recognised by the

Glover Review, among others, that access to nature was a crucial component in making our quality of life acceptable, and in improving it. In a sense, the review was echoing much of what had been said all the way back in 1945 about the prospects for British (actually, English and Welsh) National Parks as part of the post-war recovery plans. Dower's report was a key part of the deliberations which led to the National Parks and Access to the Countryside Act of 1949. It in turn led to the creation of the Countryside Commission, then English Nature and what is now called Natural England. Dower asserted that because England consisted of 'a large population living in a small island of matchless but most vulnerable beauty, it would be reckless to squander, and destroy it'. He quoted Henry Strauss MP, who had told the House of Commons three years earlier (at the height of the war) that in creating National Parks, 'there can be few national purposes which, at so modest a cost, offer so large a prospect of health-giving happiness for the people.'

When Dower made his report, the push for the industrialisation of agriculture had not begun. The population of England and Wales was less than 42 million people, and there were fewer than 49 million people in the whole of the UK. Our population is predicted to hit 70 million in the next decade.

I have already said that when we talk about protecting the countryside we must be careful about imagining how we can return it to some pristine, or even truly wild, state. In 1945, Dower and his wartime colleagues were well aware that in the UK there was almost no 'virgin country', unlike the areas preserved in Africa or America as National Parks, noting that 'there are no considerable stretches in England and Wales, and few even in the Scottish Highlands, whose landscape has not been to a significant degree modified by farming or other human uses.' They were clear that by creating National Parks, 'established farming use should be effectively maintained.' In light of how desperately so many people sought access to the natural world during the pandemic, it is sobering to read how MPs believed back then that 'without sight of the beauty of nature the spiritual power of the British people will be atrophied. The longing, too often thwarted longing, for natural beauty and the great unspoilt places is most touching and a most hopeful thing in the modern city population.' In our recent public health crisis we have learned the truth of that statement the hard way, with many organisations stating how important access to outside spaces is for maintaining good mental health.

When people talk about the government's plans to ensure that the farming community is paid for maintaining 'public goods', there is ample room for argument. Some fear it is a strategy which will only make income available to farmers who are somehow nature wardens, who must sacrifice producing food in return for being paid as 'park keepers'. It was very clear to everyone involved with the Glover Review that the functioning of the countryside as a working environment and as a natural asset was dependent on the communities within it.

The problems of traffic congestion caused by visitors, or villages being taken over by 'incomers' who visit only in the holidays, and challenges like the lack of affordable rural housing and jobs for those who keep the countryside 'alive' are pressing issues. As the Review pointed out, the preservation of our natural landscape (albeit altered by man for millennia) relies very much on human actions. The report stated that: 'It is when the beauty in nature, in geology, insect life, storms and clouds, comes together with the beauty of a hand-crafted farm gate, a Dales barn, or a shepherd's crooks, that the power of our landscapes is revealed. We believe that it is only by recognising the role of people and nature together in shaping our landscape over thousands of years, and the good and harm that they can do today, that we will once again bring our landscapes to life.'

There was a circularity to those findings. They again reflected the original conclusions made by John Dower when he wrote, 'It is above all else to farming, both to the extensive grazing of the higher open land and to the more or less intensive grazing, mowing and cropping of the lower, fully enclosed land, that the landscapes of all our potential National Parks owe the man-made element in their character; and it is to the farming communities that we must look for continuance not only of the scenic setting, but of the drama itself . . . the endless battle between man and nature – without which the finest English or Welsh scenery would lack an essential part of its charm and recreational value.'

Interestingly, the original parameters of the National

Parks didn't lay down legal protections for nature. With the crisis in the natural world, and the burgeoning threats from climate change, it's time to think again. Farmers and farming are not just part of the scenery in these nature reserves and National Parks. They work the land and have the power to make it better. If, as the Glover Review makes clear, these areas which we label as 'special' and 'beautiful' are to be part of the recovery of nature and contribute positively to our reduction in carbon emissions, while at the same time providing 'public goods' for the nation, then farmers will have a lot of work to do. As a result, the review stressed that 'fostering economic and community vitality' in the countryside was essential for these things to happen.

6
Heavy Pheasants

'To be a good gamekeeper, you need three things in your pocket: a knife, a shilling and a piece of string. Then you'll be prepared for anything, and have enough money for a pint of beer at the end of the day'. That was the advice I was given by 'Reg the stick man', the old countryman at Knepp when I was starting out. We'd go on pub crawls around Sussex, and I remember it being a particularly good stamping ground. In the summer Reg made polo sticks, and in the winter his income came from gamekeeping. You won't get much beer for a shilling these days, but he was right in many ways about being prepared. I often remember his words, because in a way they sum up the simplicity of what it means to work on the land and act as a guardian for the habitats that produce the game. In the rapidly changing world of today, the notion of what a gamekeeper is may need refreshing. However, I believe that good gamekeepers can be a key element in the restoration of nature.

In my time I have been a lowly underkeeper, a beat keeper and a head keeper. I've worked single-handedly and been part of a large team. More recently my job has entailed managing a department that includes a head keeper and underkeepers who look to me for guidance. For many people, the world of gamekeeping conjures up something outdated, and they associate it only with encouraging the killing of wild animals or, at best, looking after them until it's time for wealthy people to kill them. I don't agree. A modern gamekeeper needs many refined skills, and has to be part botanist, part naturalist, part ecologist and also a livestock manager. He or she also needs to understand a wide variety of habitats from pasture and woodland to grassland and marshes. The gamekeeper is a jack of all trades who must be aware of his surroundings and what the consequences will be of everything that happens on the land for its wildlife. It's not just the 'shootable surplus', but also the livestock, the insects and the plants, the fish and the trees. A gamekeeper understands how it all ties together.

Gamekeepers come in for some criticism, and there have been highly publicised instances of individuals illegally killing raptors or badgers to protect game bird shoots. I can honestly say that in my career I've never been asked to do anything like that. Someone once said I 'might consider going down to the pheasant enclosures' with my gun because they had seen 'a tawny owl which might bite the heads off the poults'. I did visit the young birds, but I had no intention of shooting an owl on the off chance that it might take a poult that was destined to be shot anyway.

Some years ago I was working as a loader for someone at a private shoot in the Midlands. There were eight guns, and the guests were expecting a good day. This was not a wild game shoot, so the pheasants had been reared in large numbers to ensure there would be sufficient surplus for sport. To make sure the guests maximised their enjoyment, they were provided with loaders. Our job was to make sure the shooters always had a gun ready and primed to fire once they had identified their target. In the heat of the shoot, when you're loading and passing the guns back and forth as fast as you can, you get into a rhythm and there's virtually no time to think. After one drive, I finally looked behind us and there was a carpet of dead pheasants. By the end of the day we had finished off more than twelve hundred birds. I was dumbfounded at the spectacle, and felt uncomfortable that I had been party to the slaughter. It was the sheer scale of the killing that made me uneasy.

Raveningham was known for its wild bird shoot. Once upon a time that was how it was on every estate. There was sufficient surplus wild protein for sport precisely because of the way farming was carried out. In recent decades our farmland, in general, has not delivered those surpluses. Shooting estates have had to buy in poults and feed them up for some weeks until they are large enough to fly, and with suitably impressive tail feathers to make a display as they wing through the air. This is not 'hunting' and it's worth pointing out that in the UK we don't use that word for any fowl or game shooting, reserving the term for what used to happen on horseback with hounds, or for the exotic pursuits once practised fairly widely by the wealthy on safari in Africa.

In America they talk about activities like 'duck hunting' and don't seem to make a distinction. I don't have a moral problem with shooting game when there is sufficient supply, and in the case of the UK's rampant deer population we are currently squandering a massive source of inexpensive wild protein that is very good to eat. I think that shooting sufficient game for our needs is one thing, but only if it's available because of the way we are looking after the countryside. There is clearly something questionable about commercially breeding huge numbers of pheasants and partridge – often importing them from overseas – merely to release them when it is time to provide some sport. That kind of shooting expanded massively in the 1980s when the so-called 'Big Bang' happened in the City and a lot of money started swilling around. People who associated game shooting with privilege and wealth wanted to prove their worth and achieve some kind of status by paying to experience a shoot. I never saw any of that growing up. My father used to take me rough shooting in Wiltshire, but it was just him and me, and he was a notoriously poor shot. If we took three or four wild pheasants in a day we were doing well.

No one really knows how many birds are released for game shoots in the UK. We do know that it runs into the tens of millions, and one recent estimate from a study at the University of Exeter puts the potential number at anything between fourteen and seventy million birds. The study, by Dr Joah Madden, calculates that a reasonable average of just over twenty-five million pheasants, nine million partridges and almost five million mallards are released in a year. He also calculates that of roughly ten thousand shoots in the UK, more than 90 per cent use artificially reared and released birds. Notwithstanding the estimated mean total of around forty million birds, Madden's analysis reveals that the change in the numbers of released birds since the 1960s could be an increase by a factor of nine (for pheasants), of five for mallards and a staggering two-hundredfold increase for red-legged partridges.

Numerous conservation organisations have questioned whether there may be damaging ecological consequences from this influx of reared birds into the countryside. These are, it must be said, non-native species. If we take the higher estimates for the numbers of birds released – sixty to seventy million – then they represent a biomass equal to all of the other wild birds in the country. Definitive proof that this is bad for the ecosystem is very hard to find, but one very probable effect is that the sheer number of released birds provides a significant source of extra food for predators – especially foxes and raptors – which may be artificially distorting the numbers of those species which the countryside would 'naturally' support. A contrary view holds that precisely because so many birds are being released, predator control by gamekeepers is more rigorous. Thus, in this view, shooting is a positive driver behind the creation and maintenance of woodland habitat – which clearly benefits wild species of many different kinds. There is also the perceived economic benefit of the jobs provided in the rural economy through the operation of shoots. Ecologically, opponents of game shooting and the introduction of artificially reared birds point to the potential risks of outbreaks of avian diseases and the potential for so many birds to affect nutrient levels in soils. Mark Avery, the nature writer and ornithologist, says that 'these non-native game birds go around gobbling up insects, other invertebrates and even snakes and lizards, they peck at vegetation, their droppings fertilise sensitive habitats which no farmer would be allowed to fertilise, and they provide prey and carrion that swell the populations of predators that then go on to prey on other threatened species.'

In the summer of 2020, a Rapid Evidence Assessment on the Ecological Consequences of Gamebird Releasing was conducted by Defra on behalf of Natural England and the British Association of Shooting and Conservation. Its findings were inconclusive. Noting that the release of game birds had begun in 1900 and had been increasing ever since the peak of intensive farming in the 1960s, the report used an estimate of seventy million birds (pheasants and partridges only), and calculated that 85 per cent of them were released in England. The report agrees that the releases amount to a 'perturbation of a natural ecosystem, and these effects are unlikely to be simple'. In effect, Defra found that this complex area needs further study. However, they did accept that gamekeepers were actively managing agricultural and semi-natural habitats, which led to the planting and maintenance of woodland, rough ground, ponds, hedges and game crops 'at levels higher than other land owners'. This type of land management, they suggested, would 'lead to increased numbers and diversity of plants, invertebrates and non-game invertebrates in those areas of the game shoot. Several of these species [are] of conservation interest.' In 2021, Defra announced some restrictions on the release of game birds, especially pheasants and red-legged partridge. There are limits on the numbers of birds released per hectare within what is defined as a European site (a categorisation of conservation or scientific status defined by the Conservation of Habitats and Species Regulations 2017) and regulations about recording the number of birds released within and near to these areas.

Having said that none of this is ecologically simple – as if anyone who works on the land imagined it would be – I return to my belief that the gamekeeper is a potentially powerful force for change and conservation within the countryside. Working in partnership with farmers, as they have traditionally done, gives us a very powerful weapon in making sure that our agricultural landscape is productive in every sense of the word.

I like to think that the knowledge I've gathered over a lifetime of working in the countryside has been gleaned only partly from books, and much more from talking to people 'with mud on their boots'. At Knepp it was people like 'Reg the stick man', but everywhere I've visited or worked there have been people who have been generous with their own experience and know-how. I remember that on my second day at the cattle ranch in

Western Australia I was told to go and shoe some semi-wild horses that had been running loose in the bush for six months. I had absolutely no idea how to go about it, but Buddy, one of the stockmen on the ranch, came to my rescue and taught me how to do it. You had to ride the horses you shod, and without Buddy I would have been stuck with horses that would have gone lame very quickly. He also taught me how to crack a bullwhip. In England, other gamekeepers have taught me a lot over the years, especially Charlie Mellor, head keeper to the Duke of Norfolk in Sussex. I've spent hours talking about how to create habitat for game with Robert Hall, who manages game at another large estate not far from Holkham. He understands that the whole system has to hang together, and he knows that sawfly larvae, spiders, ants and winter bird food are the crucial building blocks that underpin a healthy bird population. These men have been great influences, along with Gerald Grey, head keeper at Hilborough, also in Norfolk. Before he retired he was passionate about trying to reverse declines in wild grey partridges. In recent years Gerald has become very involved with trying to monitor and foster populations of stone curlews, which have been declining across their range in Europe for some years. Sometimes called 'Norfolk plovers', these shy birds with their conspicuously large yellow-ringed eyes are often most active at night. Gerald is now doing what he can to encourage the right conditions for the insects the birds need to thrive.

Although my father gave up farming just before I was born, we always used to visit Suffolk regularly and I absorbed a great deal from his old friend, John Horsman. When my father was a young farmer, John was a neighbour and, like him, a bachelor at the start of his career. He was a man filled with energy and wisdom. I remember him at ninety still looking and behaving much younger than his years. He was very excited to tell me about which butterflies and orchids he had seen recently on his farm. He would reel them off: painted ladies and peacocks, graylings and gatekeepers, orange tips, brimstones and many more. While my father had moved into a different world, John carried on farming, but he resisted, almost instinctively, the drive to make his farm more industrial and more intensive. He was doing 'agri-environment' at his farm near Cratfield before the term was invented. In 1985 he won the *Country Life* Farming and Wildlife Award, recognising the way he had turned his headlands (the field edges on arable land, usually wide enough for a vehicle) into grassland so as to directly benefit wild plants and animals. The following year he was the first Suffolk farmer to be given the National Silver Lapwing Award, which honoured those who demonstrated a genuine commitment to habitat and species conservation while simultaneously integrating environmental management into their farm business. The Silver Lapwing Award also required farmers to recognise the historic aspects of their own farms, and have an integrated approach to soil and water quality and the efficient use of energy. It could have been designed with John in mind. At the time he was also a key member of the Suffolk Farming Wildlife Advisory Group, an important lobbying organisation when it came to responding to bureaucratic advice which went against what good farmers knew about protecting nature. John understood that a healthy farming environment needed to be a mosaic of different habitats, ideally a mixture of grazing animals, scrub, arable fields, wildflower meadows and woodland. John's farm lay on clay and he used that to his advantage, creating a series of ponds to attract wildlife; I clearly remember seeing grass snakes swimming across the ponds. His hedges were splendidly uncut, especially on either side of a ditch, because he knew that the tall foliage created shade and kept down the growth of weeds which would have quickly blocked it.

The kind of regenerative agriculture that John knew instinctively how to maintain allows for nature to be present across a whole farm, not in just one or two small pockets. He was happy to share his recipe for creating a wildflower meadow with anyone who wanted to try it. His method was simple: sow an arable field with grass and leave it untarnished by any sprays or fertiliser. Mow it relatively late – in August – and use the cuttings for hay. Keep repeating this strategy every year and after about twenty years, he would joke, you will have a wildflower meadow. John's basic recipe is right, but the process can be accelerated considerably. What does happen is that every year the wilder species diversity tends to increase as the grass gradually impoverishes the soil. The most interesting wildflowers tend to hate richly nourished ground.

Farmers have always needed to be adaptable. Due to

the uncertainties of the weather and the fluctuations in the seasons, they can never predict how their crops will turn out. In what seems to be an increasingly volatile climate, they will need to be especially flexible just to survive. British farmers are already changing some of their customs and the way they plant to adapt to those changes. These are all part of why I believe that gamekeepers, in partnership with farmers, have a place in the future of the countryside. I think that in the course of my career I have witnessed both gamekeepers and farmers lose their way slightly. At Holkham, I know that the information we get from our gamekeepers is vital to the way we manage different areas of the farmland, because they see which birds and animals are present in which areas on a daily basis.

Perhaps in the modern countryside we need to start thinking about gamekeepers in a different way. They can be key contributors to the provision of the 'public goods' envisioned in the government's 25-year Environment Plan. Those aims are relatively simple: cleaner air and water, plants and animals which thrive and a greener country. The plan calls for 75 per cent of protected terrestrial and freshwater protected sites to be in favourable condition – securing their wildlife value. Outside the protected network, the plan aims to create or restore 500,000 hectares of wildlife-rich habitat and aims to increase woodland in England to achieve 12 per cent cover by 2060. In my view, these targets will be achieved more efficiently with the help of good gamekeepers.

If you consider how gamekeepers already manage

wetlands for breeding birds, storing water and creating healthy soil by encouraging biodiversity, then there's no conflict. Gamekeepers are advocates for planting crops and plants that provide winter feed for birds – as we have done at Great Farm – and they know that planting single-species crops doesn't have the same positive results. Gamekeepers understand that specific seeds are needed at certain times in order for different bird species to successfully rear their young. It therefore seems sensible to re-imagine the role of the gamekeeper. They might be called 'wardens' or 'rangers', titles that combine both the targeting of certain species in order to provide surplus protein, but also to make sure that the carrying capacity of the land is kept in balance. Whatever we choose to call them, they are essential custodians of the landscape, and in many cases their eyes and ears are among the best sources of information on habitat health.

Field realignment at Great Farm makes planting, spraying and harvesting more efficient.

At Brick O'Longs the field has been managed in several different ways to provide a range of benefits to wildlife throughout the year. This includes sections for ground nesting birds, plants that benefit pollinators and seed-bearing crops for winter feed.

The heavily farmed fields at Great Farm showed signs of serious erosion. The stumpy, gap-filled hedge provided no shelter, no food and no nesting habitat for farmland birds. Many fields are left like this through the winter, effectively forming ecological deserts in the English countryside.

At Brick O'Longs the earth was fissured with great ruts and chasms after being left exposed over several months during winter.

Field margin at Tinkers Hole in early summer.

Allowing the hedge to grow saves time and money, and provides a host of ecosystem services but has no impact on food production.

In the middle of farmland, at the old decoy wood, the mixture of water, old trees and dense vegetation provides an ideal nesting habitat for spoonbills and cormorants.

Great Farm field edge planted with wild bird seed which provides food for pollinators in summer and birds in winter.

Six-spot burnet moth (*Zygaena filipendulae*) on small scabious (*Scabiosa columbaria*)

The chalkhill blue (*Polyommatus coridon*) survives on calcareous grassland and relies on the presence of horseshoe vetch (*Hippocrepis comosa*) as food.

Pyramidal orchids (*Anacamptis pyramidalis*) prefer grassland habitats, and have returned in large numbers, especially at The Burrows.

The bee orchid is another grassland favourite less common now, but encouraged by cutting the grass at the right time of year.

Delicate bladder campion (*Silene vulgaris*) is a meadow species, sometimes called 'Maiden's Tears', and a favourite with pollinators.

A typically bad example of a commonly seen brutally shorn hedge which is of little use to man or beast.

With careful management, grazing animals like these Belted Galloways can contribute positively to grassland habitats and bring ecological benefits to other species.

Areas of shallow water are a thriving habitat for insects that will support nesting waders and other birds. It's a perfect example of 'edge' where water, grass and mud all meet to create an abundance of life.

At Holkham we have maintained a healthy breeding population of natterjack toads.

In summer the high hedge is filled with diverse plants for pollinators and provides a nesting habitat for birds like yellowhammers and linnets.

In late summer and autumn the hedge is covered in fruit and berries.

The joyous sight of the returning pink-footed geese in winter.

At The Burrows, sensitively winter grazed meadows provide food for sheep and become a great mass of insect friendly flowers in spring and summer.

A cultivated area around the crops where annuals like poppies flourish, providing sparse cover for declining farmland residents like grey partridges (*Perdix perdix*) and skylarks (*Alauda arvensis*).

Field 162
April / May

As dusk approaches, and the marsh birds settle for the night, the sounds of the calling toads begin to dominate. Thousands of male voices rise like a choir to make a gentle rolling *urr-urr-urrk* that echoes across this tussocky field close to the dunes. A series of small ponds lie in the lee of the dunes. About 150 yards across the field there is a much bigger pond, hidden by thick clumps of rushes. Lapwings nest in the field in springtime, and during the lengthening days birdwatchers peer across the fields from a vantage point on the raised sea wall that gives views towards the Overy marshes.

West of the pinewoods, some of the little ponds are barely larger than a decent-sized Jacuzzi, and just a foot or so deep in the middle. In late April we start checking them every day, looking for the tell-tale signs that the natterjack toads have been laying their eggs. The water is very still here, warming in the spring sunshine and with the tint of cold tea; the ponds are not yet thick with vegetation. The toads need the warmth to hatch their eggs, which hang in the water like jet-black pearls, in double rows at first, but then after a day or two separating into long single strings.

These small ponds are found at the northern edge of

Field 162. I wish it had a more romantic or at least a historic-sounding name, but it doesn't. The site is protected, and so are the toads which are now only found in about sixty locations in Britain. Apart from the colonies in East Anglia, there are also scattered populations in the west on Merseyside, on the coast of Cumbria and the Scottish Solway, with a few also in Hampshire and Surrey.

Cattle occasionally graze the field, and now, due to climate change, we have become accustomed to the sight of great and little egrets following the beasts through the long grass. I remember when we first started seeing the birds in Norfolk in the 1990s, and it was big news when little egrets first bred on the south coast. Now they are part of the landscape. The much larger great white egrets (about the size of a grey heron), seem to be one of the species which is doing well from the changing climate. Fifty years ago there were just a few hundred pairs breeding in Europe and now their population is in the order of thousands; Along the Norfolk coast we see them year-round. For all that the birders love seeing them, I do wonder whether they may eventually have an impact on the natterjack population, as frogs and toads are definitely on their menu.

I often spot natterjacks when I'm out walking in Field 162. It's known that they need areas of short grass where they can easily move around and find prey, including slugs, snails and insects. They seem to make the journey back and forth from the sand dunes to the large pond easily. When the male and female toads find one another, their mating activity – called amplexus – can be brief, or it might last for several hours. The male, usually the smaller of the two, clasps his front legs around the female's shoulders and under the 'armpits', hanging on firmly using the nuptial pads (patches of rough skin) that develop on their toes during the breeding season and then subside. Toads, like other amphibians, fertilise their eggs externally as they are laid, and the female will spawn as many as seven thousand. which can hatch into tadpoles in as little as a week. The tadpoles have a precarious existence and are potential prey for newts, leeches, diving beetles and dragonfly nymphs, as well as for herons and egrets.

The larger pond, which is really a marsh, extends over an area that can be anything from just under half a hectare to about three times the size depending on the water levels. This area is really just a depression in the field which was created twenty years ago by a previous reserve manager. He hoped it would attract avocets. To create the habitat he bulldozed a scrape to make a wetland, but when he moved away from Holkham the area was not actively maintained. Over the years it became overgrown with vegetation and colonised by plants like bulrushes and the ever-present Juncus. For many years no one knew that the natterjacks had moved into this larger area, which is in the middle of a field used for cattle grazing, as it was assumed they were always found close to the dunes. In winter the toads hide away in burrows and are known to make use of rabbit holes in which to hibernate.

Historically, cattle were kept away from the small ponds near to the dunes, on the grounds that they might trample the toads or remove the vegetation in which they sheltered. In 2019 the ponds had very low water levels and the number of egg strings was worryingly low. Then we noticed that in spite of a seeming absence of amphibians at the small dune ponds, the mating chorus was audible – but coming from the other side of the field. Natterjack mating calls can easily travel two kilometres or more. So, we investigated the marsh pond, the former avocet scrape, and discovered there natterjacks in astonishing numbers.

One of the changes we made at Overy marshes was to increase the number of water-management structures fed by springs on the southern side of Holkham. The water is good quality, and has come from the aquifer, filtered through sandy loam and chalk, and we can easily adjust water levels in many different areas. The other change on Field 162 was that we allowed a small herd of Belted Galloways to graze the field over winter. It costs approximately £20 per week to house a cow, so putting twenty animals out on the marsh for sixteen weeks saves the farm £6,400 annually. The year after the cattle were allowed to lightly graze the area, we noticed that the number of toadlets which hatched was at record levels; the cows had worn a series of defined pathways around the edge of the marsh. Once the baby toads hatched, many of them seemed to actively seek out cattle hoofprints in the soft mud as hiding places. Studies of other natterjack populations have found that they need these areas of short sward as part of their habitat, as this is where they most easily find insects to eat. The shorter grass and trampled ground also allows the toads to move more easily around the area. The cows are taken off the marsh ponds during the natterjack mating time. They are also taken off in November for worming, and kept off the land while they are still excreting any traces of their medicine.

Having now seen the beneficial effects of the cattle, this winter we are also experimenting with ponies on Field 162 to see what happens. We used Koniks, a hardy breed from Poland which are semi-wild. The plan now is to keep cattle off Field 162 for most of the year and winter-graze a portion of it with four mares. The Koniks have been elsewhere on the estate since they were acquired about five years ago, before I started here, and put to work at Castle Acre, a wet boggy area at the southern edge of Holkham Estate. The hope is that they will control the invasive and dominant plants such as brambles, willow scrub and rank grass on the northerly end of the field close to the dunes. The beauty of Koniks is that they are known to tolerate wet marshy areas better than almost any other horse breed. They were used first by the National Trust to graze Wicken Fen reserve in Cambridgeshire, and have proved suitable in many areas of lowland Europe where there are wetlands. A herd was established to rehabilitate wetlands at the Oostvaardersplassen reserve in the Netherlands in the early 1980s. At around 400kgs, the ponies are lighter in weight than cattle and their smaller hooves can trample coarser types of vegetation without as much damage to wild-flower seedlings.

When assessing the impact of livestock on a natural ecosystem, whether they be horses, cattle, sheep or, indeed pigs, it's important not to automatically assume that grazing or trampling the ground is a bad thing that should never be allowed. Grassland, as the countryside historian Oliver Rackham points out, 'is easily spoilt by mowing instead of grazing, or by cattle instead of sheep'. Ruminants are a natural part of many healthy diverse ecosystems – although in nature they keep moving, either in search of fresh growth or because they are evading predators. The African savannah would not function the way it does without the movement of wildebeest, impala or other grazing species. It's been known for many years, for example, that if elephants are removed from the bush then other animals suffer. The elephants knock down saplings and create space for other plants. The same thing happens in the British countryside, where cattle, sheep and deer create space for other species. The fact that wild grazing animals are constantly moving – always alert for something that will chase them – means they don't overconsume any single type of vegetation. Without elephants, or even elk or wildebeest, we have only the wild deer, the semi-wild ponies or the domesticated cattle and sheep to fill their ecological gap. So when people say 'cattle are bad', I repeat the saying: 'It's not the cow, it's the how'.

There is a special tranquillity to being out near the toad ponds on a late-spring evening. If I am fortunate enough to catch sight of a mating pair at the edge of the marsh, they are usually oblivious to my presence, hanging there in their coupled state with just their eyes and their nostrils above the water. If my shadow falls on them, or they detect movement, they will swiftly dive into shadowy reedbeds, leaving a trail of tiny bubbles in their wake. At least, the female will dive – her male passenger has no choice in the matter. All around me, the whirring toad song cutting across the marsh is comforting, a kind of natural white noise, a signifier of plenty and the arrival of the new generation.

7
Fine Wine and Turnips

On rising ground, 750 yards north of Holkham Hall, a classical stone column towers into the air over the surrounding parkland. It commemorates the life and work of the 1st Earl of Leicester, Thomas William Coke (1754–1842). Trees were felled when it was first unveiled in 1850 so that it could become a landmark for mariners at sea. The monument looks back towards the great house, out over the rolling parkland and across the lake. The 140-foot-high Corinthian column, like others of its kind, is itself fairly standard and draws on elements from the slightly earlier columns erected in London for Lord Nelson and Sir Christopher Wren. It's surmounted not by a human figure, but by a coronet held up by four bulls and topped by a lantern holding a sheaf of wheat. At its base is a large plinth, or pedestal, ornamented by sculptures of Southdown sheep, a Devon ox, a plough and a seed drill. Each of them has its own motto: 'Small in size, but great in value' (under the sheep); 'Breeding in all its branches' (under the bull); 'The improvement of agriculture' (the seed drill); and, finally, 'Live and let live' (beneath the plough). It always seems to me that those precepts and ideas have stood the test of time. More than that, they have taken on a great relevance today. At Holkham, the concepts still hold true even though they referred to very specific acts and policies carried out by the Earl of Leicester in the eighteenth century. More broadly, these mottoes embody the disparate elements that are necessary for a farming system to work well, and that we would do well to emulate in modern agricultural practice.

The Coke monument is no dry historical relic, nor is it simply an expensive commemoration of a wealthy landowner (although it was built at a cost of more than £5,000 – more than £600,000 today – funded by donations from more than a thousand local gentry and Holkham tenants). The elements within the design and construction are highly specific. The proud ox and the sheep show that Thomas William Coke, the 1st Earl of Leicester, was interested in improving his livestock through selective breeding. The plough and the seed drill demonstrate his willingness to try out new technology and apply the latest principles to farming methods. The wheat sheaf is an emblem of food production which encapsulates his renown within the history of English farming, but it was chosen because Coke is credited with dramatically increasing the amount of wheat grown in the county.

Holkham's connection to agricultural experimentation and increasing productivity are all around. In one of the cool, tiled entrances to the Hall, there is a dusty glass display cabinet beside the doorway. Inside, totally desiccated and bleached by time, is a collection of ancient sheaves from across the globe, evidence of how the breeding and selection of crops has been undertaken here over the centuries. There is Peruvian Barley and Hudson Bay Grass, Sprat Barley, Black Winter Bere and Portuguese Rye. Wood Meadow Grass sits beside Confused Fescue along with Marram, False Oat Grass and Marsh Bent. The litany of intriguing common names is evocative, even on a dull autumn day, with Yellow Oat Grass and Creeping Soft Grass, Yorkshire Fog, Crested Dog's Tail and Red Fescue conjuring the scents of high summer and ruminating livestock.

The Duke of Sussex claimed that Coke's achievements were all the more remarkable because the land he acquired to build his farms previously amounted to 'little more than a rabbit warren'. That may be a bit of aristocratic hyperbole, because from the Middle Ages this part of Norfolk was known as the 'Sheep-Corn' region, an arc of fertile sandy soil extending from the south-west to the north-east of the county. Sheep were known to 'tathe' the land, a unique regional term for fertilising the soil with their droppings. Notwithstanding Sussex's flattery, there is no doubt that many of the farming methods adopted by Coke – and which he made his tenants implement – resulted in increased efficiency, better yields and general improvements to the soil. Many of these improvements were already at the forefront of what was happening in English agriculture at the time. Perhaps by good luck, Coke's lifetime coincided with the so-called British Agricultural Revolution, when production rose to new levels and a combination of technology and improved farming techniques allowed a surge in the population, which almost doubled – from just over five million in 1700 to nearly ten million in 1800. Improvements in diet and animal husbandry alongside selective breeding meant that dairy cows went from producing around 350 to 420 gallons of milk per year, while bulls for slaughter went from 400 to over 600 pounds.

Quite simply, the dramatic changes of the Agricultural Revolution meant that there was more food available. Now, with our world population still rising and projected to reach 9 billion by the middle of this century, how we produce enough food is set to be a major challenge. It potentially requires another agricultural revolution, especially if increased food production is to be achieved in a sustainable, healthy way. We can grow enough food for an enormous population, but we must do it in a way that leaves space for nature.

While we confront this major challenge, the UK has made a legal commitment to reach what it calls 'net zero' greenhouse gas emissions by 2050. It's Britain's contribution to trying to stop catastrophic global warming, the evidence and signs of which are overwhelmingly clear. How we use our farmland is a key part of the battle to stave off that catastrophe, and it's also a serious weapon in mitigating some of the really negative effects of that climate change – whether that's flooding, soil erosion or a decline in food production. The part that farming has to play in these net zero commitments is enshrined in the government's 'Agricultural Transition Plan', as published in late

2020. This post-Brexit strategy was partially informed by the consultation exercise carried out by the government to produce yet another weighty piece of research: *Health and Harmony: the future for food, farming and the environment in a Green Brexit* (February 2018). This was a rigorous examination of the harm caused by the Common Agricultural Policy, and somewhat trounced the idea that no one had thought about the consequences for food production of leaving the EU – taking into account more than 40,000 individual responses and almost 130,000 signatures to several petitions relating to the consultations. It also built on numerous countrywide round-tables, taking evidence from land managers and stakeholders, and the consultation fed into the drawing up of the 2020 Agriculture Act.

Radical, almost unimaginable changes within farming are not a modern phenomenon. In the eighteenth century, greenhouse gases had not been identified and the Industrial Revolution was only just beginning. However, big changes in agrarian methods were part of the process of industrialisation, and Thomas William Coke was not alone in adopting new farming methods. A generation earlier, another local landowner, Viscount Charles Townsend, had promoted what became known as the Norfolk System. Previously, farmers had left the land fallow – or unproductive – for a year every three or four years in order for the soil to recover and replenish its nutrients. The fallow period was an antidote for what farmers termed 'land that is plough sick'. The Norfolk System avoided this through crop rotation. Townsend wasn't the first to discover the benefits of crop rotation, but he promulgated it among his circle of aristocratic landholders, who were always interested in making profits. Earning himself the nickname of 'Turnip Townsend', he would plant a series of crops – wheat, turnips, barley and clover – in successive years. This 'four-course rotation', sometimes called 'corn, turnips, corn, grass', was described by his contemporary, Lord Ernle, as 'farming in a circle, which, unlike arguing, proved a productive process'.

The popularisation of the turnip, although grown over small areas (rather than as a whole-field crop) in England for centuries, was at first regarded with suspicion as a Hanoverian import, and Townsend was mocked for his experimentation. However, as another contemporary admitted, 'He none the less went on with his work, and in a few years changed a poor fruitful district into one of the most thriving in the kingdom. All the interest of the preceding generation had been in hunting – their talk, nothing but horses and dogs: the talk of the present one was all manure and drainage, rotation of crops, clover, lucerne grass and field turnips.'

Coke embraced the new fashion wholeheartedly and on Holkham farms the requirement for five-, six- and even seven-course rotations became more common. Leases stated that 'the tenant shall not sow any more than two successive Crops of any sort of Corn Grain or Pulse . . . and one of such two successive Crops shall always be of Pease or Vetches, so that two white straw Crops of Barley, Wheat, Oats or Rye may not in any case follow immediately after one another.'

Coke knew that the alternation of deep and shallow-rooted crops helped improve the structure of the soil. Turnips, with their deep roots, also provide useful winter food for animals and their deeper roots help break up the soil, thus relieving compaction. English farmers had traditionally (before the Inclosure Acts) simply alternated fallow fields with crop growing. Where three fields were available, they could leave one fallow every third year, and plant legumes in the second field to improve soil nutrition after a year growing wheat or barley. Anything from a fifth to a third of arable land had to be left fallow in any one year. The importance of the new rotation was that it included both fodder crops (which nourish the soil by absorbing nitrogen from the atmosphere and 'fixing' it in the ground) and grazing, allowing the farmer to breed livestock all year round. In general, grain crops deplete the soil of nitrogen, and without modern fertilisers they could not be grown repeatedly without declining severely in quality, and subsequently delivering poorer yields. Coke adopted the Norfolk rotation enthusiastically, and made it a condition of his tenant farmers' leases that they abstain from planting what he termed 'white crops' (grains) on the same land more than twice in succession. What's interesting to me is that the revolutionary implementation of strict rotations allowed arable farming to be better integrated with the keeping of livestock. Winter roots like the turnips provided fodder for greater numbers of animals, which in turn provided more manure. It's the circular nature of all things, and a recognition of their interdependence, that we are trying to foster with modern regenerative farming.

Coke also granted long leases of nineteen or twenty-one years, which, according to Lord Ernle, 'guarded against the mischief of a long unrestricted tenancy by covenants regulating the course of high class cultivation'. In addition to crop rotation, the leases stipulated that land had to be returned to pasture for two years at the end of each four-year rotation. When a grass ley was planted, the leases stipulated that it could only be mown once, or pastured (grazed and fertilised) by sheep or cattle. Grass was never planted on its own; the leases ensured that one acre was to be sown 'with 12lbs of grass seed mixed with clover, and 1 peck of rye grass'.

To make space for more livestock, Coke also made good use of the controversial practice of 'Inclosure', where formerly common land was fenced in for the exclusive use of an individual. The process had begun in medieval times and increased in the Tudor period before reaching a peak in the nineteenth century. Between 1604 and the early twentieth century, some five thousand Inclosure Acts were passed in Parliament covering almost 7 million acres. The Acts were another driving force sending people into the towns and cities, thereby contributing to the Industrial Revolution.

Whatever the rights and wrongs of the changes, it led to more efficient farming, higher levels of food production and ended the communal nature of Norfolk livestock rearing, which involved a complex 'foldcourse' system to allow tenants and landowners to share grazing at certain times of the year. It also allocated certain areas for planting by individual village communities. One of the problems with common land was that cover crops could not be planted and left to help heal the soil, as they would prevent other people's livestock using the land for grazing.

The enclosures also left a legacy of hedgerows which subsequently became a defining feature of the English countryside. Often portrayed as a process which took away land from the common man, the fencing-in process was part of the modernisation and streamlining of the English agricultural system. The familiar hedgerow, creating the distinctive patchwork of fields that we think of as 'the countryside', is, in environmental terms, a fairly recent innovation, but they especially suit birds like linnets and partridges, which favour them as shelter but require the adjacent fields for foraging.

The 'open field system' that dominated parts of Britain before the Inclosures is still found in parts of Eastern Europe, where the variety of bird life and numbers of individual species are often healthier than in the UK. As the more open, more diverse landscape here was altered by the planting of hedges as fences, the longer-term consequence was that they became a vital buffer to the steady eradication of the natural thorny scrub on which so many of our birds depend. Here in Norfolk, many of the existing hedgerows are relatively poor as habitat as they are predominantly hawthorn and not much else.

Thankfully in the last few years the emphasis – and government prescription – has been on planting new hedges which are species rich. In thirty or forty years I think we'll see many very different and varied types of hedgerows around the countryside.

In 1816, Coke's plans were boosted by the appointment of a man named Francis Blaikie as his steward and land agent. He would become a crucial part of the workings of Holkham for the next sixteen years, even proffering investment advice to his employer at times and reining in some of Coke's extravagances. Blaikie's arrival at Holkham coincided with what has become known as the 'Year without Summer', when temperatures throughout the northern hemisphere were several degrees below normal, due to the massive ash cloud generated by the severe eruption of Mount Tambora in Indonesia the previous year. In what some might see as a preview to the kinds of disruptions to the natural seasonal cycles that we may face with climate change, crops failed across Europe. 'Bread riots' occurred in many countries. In England, the price of wheat soared by more than a third in the space of a month, and the farming communities of East Anglia were especially badly affected. Confronted by an angry mob set on storming Holkham, Blaikie met them on horseback at the gates and told them forcefully that Lord Leicester would give them whatever they needed if only they would ask calmly, 'like honest men and not howling children'. Blaikie then ordered that the hungry crowd be given slaughtered sheep and whatever food they needed from the estate stores, averting a disaster. Coke rewarded him with the gift of a silver snuff box.

The son of a farmer, Blaikie had trained at Kew Gardens and already had considerable experience, having worked as land agent for the Earl of Chesterfield in Derbyshire until his death. Known for his mechanical skills, Blaikie had personally invented, or improved the design of several farming tools, working with blacksmiths to modify implements like spades and loppers to make them more efficient for particular jobs. Before coming to Holkham, he had published successful pamphlets on the planting of trees and how to cultivate Swedish turnips. He later wrote *Observations on the Conversion of Arable Land into Pasture* and *A Treatise on the Management of Hedges*. Under Coke's direction, he oversaw the planting of around two million trees on the estate. On hedges, he recommended cutting the stems with an upward motion rather than 'hacking downward . . . which leaves the stubs shattered . . . so that the wet descends into the crown of the roots and produces canker. The young shoots which spring from such stubs may be numerous, but they will be puny and feeble, scarce able to support their own weight and the first high wind or fall of snow, will break them down.' Blaikie also notes that there was a particularly violent method of hedge-cutting employed by some farmers in Norfolk called 'buck-stalling', which involving cutting the hedges short and flailing the wood until it shattered. He believed that the technique originated in the days of William the Conqueror, perhaps to encourage new green shoots to benefit deer in the areas where the King wanted to go hunting. He goes on: 'There are still to be found some modern Goths who practise buck-stalling, but these are men who have either not given the subject a thought, or are blinded by prejudice.'

It's fascinating that nearly three centuries ago at Holkham they were debating how hedges shouldn't be treated too harshly. This chimes absolutely with my own belief in the value of treating hedges with a more subtle approach, respecting their complexity and high biodiversity value. The idea of allowing mixed planting was also part of their strategy, with the *Journal of the Royal Agricultural Society* commending Coke in 1795 for allowing 'the neighbouring poor to plant potatoes among his young trees for two to three years, which keeps his land clean and saves hoeing'. One of his obsessions was improving the quality of the soil on his own farmland and that of his tenants. He planted large quantities of sainfoin (*Onobrychis viciifolia* – a perennial legume sometimes called 'holy hay'), which fixes nitrogen in soil efficiently, as well as making it more fertile by increasing the ability of the soil to hold phosphates. It's especially attractive to pollinating insects and acts as a natural anti-worming agent for livestock. It's also good for getting livestock to put on weight rapidly and, although not relevant for Coke, we now know that it causes livestock to produce fewer methane emissions. Coke also planted Cocksfoot Grass (*Dactylis glomerata*), another good hay crop which improved soil health. Coke was able to more than triple the number of sheep he kept on his estate through a combination of selective livestock breeding and by growing better hay grasses. According to the Royal Agricultural Society, 'the quantity of sheep raised was surprising, and the land renovated by lying two or three years under grass, and enriched by the manure derived from the sheep. Leicester also turned all the ducks he could get onto a field to kill the "black canker", a fly larva feeding on the turnips. 400 ducks cleared the whole field in five days.'

Coke of Norfolk, as Leicester was usually known, believed that manure was a critical element in good agriculture. With crop rotations came more productive land, and this in turn supported more livestock. The animals, in turn, produced more manure, which was spread onto other arable parts of the farm. High quantities of manure, made available by the extra livestock supported by land made more productive through the new rotation system, were applied to his estate farms.

With regards to sheep in particular, Coke experimented with cross-breeding the existing Norfolk black-faced sheep with New Leicesters (bred in Leicestershire, and not named after Coke as Lord Leicester). For a time, he even kept merinos imported from Spain. The New Leicesters (also called Dishleys) embodied his motto: 'Small in size, but great in value'. They were an early fattening breed developed by Coke's contemporary, Robert Bakewell (1725–1795), who was also known for his innovative approach to farming, especially of livestock. He was one of the first to experiment with what he called 'floating meadows' – farmland that was carefully irrigated and flooded in the autumn to produce an early flush of grass which could be used at the very beginning of spring to feed his hungry winter livestock. Although he held only a few hundred acres and did not possess a grand estate like Holkham, Bakewell, like Coke, travelled widely around England and the continent, making notes on things he liked or could be adapted to improve his own farm. He was one of the first farmers to rigorously segregate his male and female stock so as to avoid accidental breeding. His stumpy-legged New Leicester sheep gained weight more quickly than other breeds, and the meat had a high fat content, which was a valued quality at the time due to the cost of tallow. Coke crossed them with the taller, fitter Norfolk breed and for a time had high hopes they would prove superior. However, the New Leicesters had to be slaughtered before they reached the age of two, otherwise they became very fat. When someone complained that the meat of the New Leicesters was of inferior quality, Bakewell replied, 'I do not breed mutton for gentlemen, but for the general public.' He also developed an important herd of Longhorn cattle, and charged stud fees for his horses, bulls and rams which were up to ten times the norm.

As well as adding manure to the land, the soil on the Holkham estate farms was also improved by a concerted programme of 'marling', a way of improving farming soil by mixing it with calcium carbonate from limestone combined with clay (which improves the soil's ability to hold water). Marling, sometimes called 'mineral manure', has the effect of reducing the acidity of the soil, and in Norfolk the term also applies to the application of lime, white chalk or sandy loam, as well as several other naturally occurring composites. It's an ancient technique, and in Norfolk accounts for the presence of a large number of pits and ponds across the county from where the marl was excavated. In Coke's time, marling became a significant activity, as described in an account of 1768 by Arthur Young: 'All the country from Holkham to Houghton was a wild sheep-walk before the spirit of improvement seized the inhabitants. What has wrought these vast improvements is the marling; for under the whole country run veins of a very rich soapy kind, which they dig up and spread upon the old sheep-walks.'

Aside from the very obvious symbols of the animals and crops grown at Holkham, the Coke monument also bears a series of bas-reliefs showing Lord Leicester in the company of Blaikie presiding over the signing of new leases for tenant farmers. Another bas-relief shows Coke overseeing new irrigation works, representing the significant reclamation – some 400 acres – of land from the sea on the northward edges of the marshes. He is also depicted at one of his famous 'Shearings', the annual event which perhaps did most to spread the word about his own farming experiments at Holkham. Nominally held at the time of sheep shearing, and at first including only local farmers, the event gradually developed into a three-day party when other landowners, farmers and tenants would gather at Holkham to discuss the latest farming practices and tour the farms. Coke, keen to 'collect around him practical men', hosted the event from 1800 to 1821. Towards the end, he was providing dinner for up to seven hundred people and allowing as many as eighty to stay at the Hall, requiring the hiring of extra beds. Estate records note that the guests included 'dukes, Members of Parliament, Americans and liberal men from all parts of the kingdom and from abroad'. Coke's hospitality was very liberal, with the 1818 accounts showing that the guests consumed 996 bottles of port, sherry and wine over the three-day event. Francis Blaikie sometimes despaired of Lord Leicester's extravagance, but at the same time his investment in new land acquisitions and the improvements in farming techniques saw his estate incomes rise considerably. By 1816, Coke was earning almost £26,000 annually from land that in 1776 had brought in less than half that amount.

While historians debate just how crucial Coke was as an individual innovator in many of these agricultural achievements, he doubtless made Holkham a centre for discussion. Perhaps his most notable success was in getting people to visit the estate and talk about the things that were happening in agriculture, and from there spread the word more widely about what the future of farming could be.

8
On the Edge

Everyone who works with me knows that I never get tired of shouting out one particular word: 'EDGE!' It's probably my favourite topic, even more important than talking about the need for healthy hedgerows and lamenting the way some landowners insist on making them look neat at the expense of all kinds of birds and insects. Oliver Rackham put it succinctly in the 1980s, as he railed against the obsession with neatness that accompanied the intensification of farming. He lamented 'the little, often unconscious vandalisms that hate what is tangled and unpredictable and create nothing'. *Edge*, and *edges* are fundamental.

For me, an edge is where something ends and something new begins. It can be a tiny thing – a depression in the mud, or a cliff on the edge of the ocean. The edge is where two habitats collide. A change, even in the seemingly hard, immovable parts of the physical materials of the land, means flux. To me, the clearest sign of how nature works is when you get a change in the landscape. A hedgerow marks the edge of a field and is often a foraging area. It creates things: a barrier, a windbreak, a shelter, a change in temperature. A stream marks the divide between marshland and crops. A beach is the edge of two worlds: the sea and the land. The scrubland at the edge of the pine forest is where the birds and the butterflies and the dragonflies are found in high numbers. It's where the visible changes happen, and where the species congregate to take advantage of those changes. Perhaps it will mean different things to a biologist or to a farmer, to a woodsman or a gamekeeper, but for me it's a vital concept.

When it comes to managing and creating habitat, a lot of the time it really is *all about the edge*. When it comes to habitat restoration and making farming friendlier to wildlife, it's as important as studying the way tectonic plates rub alongside one another for a geologist. It's where the contrasting elements in the land mix. The same thing happens in the ocean, and fishermen know that the places where cold and warm water meet are often most abundant in marine life. These so called 'liminal zones' are places of movement, and often of fertility. Edge is about having numerous things going on, adjacent to one another but which remain different. It could be something as simple as dry mud next to wet mud, or the way a hay meadow sits alongside a cultivated crop, or even the habitat contrast between a road and a field. One thing stops and something else begins. It creates competition, and opportunity.

For most visitors, the easiest way to reach Holkham beach is along a long avenue lined with poplars known as Lady Anne's Drive. When it was first built in 1823, this was a purely private road allowing the family to reach Holkham Gap, a distinctive break in the sand dunes that gives onto the open beach. Now the driveway is used as the main car-parking area for beachgoers and it's popular with riders as a place to park their horse-boxes before riding on the sands. Originally it was planted as an avenue lined with holm oaks (Quercus ilex), until they were swept away by a dramatic flood in 1953. There are holm oaks (*Quercus ilex*) elsewhere at Holkham. They are an evergreen Mediterranean species that isn't indigenous to England. Sometimes called 'holly oaks', the first specimens are thought to have been brought into Devon in 1803. For many years, the young branches from the pruned holm oaks at Holkham have been sent to Regent's Park as a treat for the giraffes at London Zoo.

The Holkham oaks that survive today arrived by happy accident. They were grown from acorns that were found among the branches and leaves used as cushioning in the packing cases by the 1st Earl of Leicester, Thomas Coke, to protect the artwork and statuary he sent back home during his European Grand Tour. Coke made the journey to the Continent in 1712 when he was just fifteen and stayed away for six years. It sparked a deep knowledge of art, and he went on a buying spree for artefacts in Florence and Rome, where he stayed to study drawing and architecture. He also spent time in Sicily, Malta, Vienna, Prague, Berlin and Amsterdam – just some of the places in which he studied and caroused. A keen student, he sent back prodigious amounts of art, including sculptures, paintings and books. He frequently had to write to his guardians in England to ask for extra funds because, for example, 'being in Rome, and seeing such a number of fine things, I must confess that I could not hinder myself from making free with more money than my allowance.' The idea that these oaks – which would live for almost 250 years on Lady Anne's Drive – had arrived by mere accident as packaging materials, seems somehow typical of how we, as human beings, can create landscapes incidentally, as by-products of other activities. Man has done many other things by fate or chance, often much bigger in their impact on the rest of the planet. As caretakers of the countryside we are often unaware of the long-term consequences, whether good or bad, of our actions. One local example is Lady Anne's Drive, which was built across land that had originally been saltmarsh and reclaimed through successive embankments built between 1659 and 1859.

On 31 January 1953, Holkham was one of the casualties of the great North Sea flood. If anything was a reminder of the potential power of nature, this was it. Following high winds at sea, a surge tide hit the coasts of East Anglia, the Netherlands and Belgium. The Netherlands were catastrophically swamped, with more than 30,000 farm animals drowned and 50,000 buildings suffering flood damage. More than 2,500 people died, the majority in the Netherlands, but among them there were 307 fatalities in Norfolk, Suffolk, Essex and Lincolnshire. Around 230 people drowned at sea, more than half of them passengers aboard a Northern Irish ferry, the *Princess Victoria*. In all, more than 250 square miles of England's east coast was flooded. Here, the sea banks at Wells-next-the-Sea and Burnham Overy were washed away with the tide as it surged through the Holkham Gap and swamped the marshes, taking a railway line with it. The tide came so far inland that the brick wall around the parkland was breached in several places and thirty acres of the pinewoods were destroyed. Trees which survived the initial tidal wave died over the coming months as they faced a twice-daily drenching by seawater that reached ten feet up their trunks. This is what happens when the equilibrium between edges becomes blurred.

In recent years, the avenue of oaks at Lady Anne's Drive has been replaced with non-native poplars, which have now grown tall and spindly with no bushy undergrowth. The new plan is to remove them, in stages, and plant a mixture of native trees and shrubs, including hawthorn, hazel and field maple. This will create a screen, hiding the parked vehicles in the landscape, but also providing a much better habitat for various types of insects and a broader range of birds, especially fieldfares, redwings and thrushes. The new planting will also sequester more carbon in the longer term.

Small changes, as I have said before, can bring big benefits. To the east of Lady Anne's Drive there are two marshy fields. I call them 'Berties' and 'Micklefleet' (originally a creek which connected to Wells-next-the-Sea and was once navigable by small boats). Berties lies to the east – the right-hand side as you head to the beach – of Lady Anne's Drive, with Micklefleet directly to the south (nearer to the A149). As soon as I arrived at Holkham it seemed obvious to me that these two areas could be transformed quite rapidly into better natural habitats. The winter of 2019 was quite dry, and on 19 January we used a digger to build several small bunds (earth walls) in the low areas of Berties. These were the original salt-water creeks which had deliberately been kept dry for many years to make farming access easier. Using an old sluice gate which had been long unused, we allowed water to flow into the low areas of the dyke. Within a few hours water began seeping into the field, and after two days all of the low ditches were filled with water. We then did the same in Micklefleet. Exactly four weeks later to the day we saw spoonbills for the first time ever in Berties. In the same week Andy Bloomfield, our longest-serving nature warden, was reporting avocets, redshanks, coots and mallards. And, notably, lapwings. To our knowledge, this field had not, in living memory, ever had breeding lapwing on it.

By 19 June, four months after we first saw the lapwings on Berties, we had newly hatched chicks. This site, which was in easy view of the visitor's centre at the end of Lady Anne's Drive but off limits to people, dogs and other natural predators, had thus become a place where lapwings were able to successfully rear chicks, not simply lay eggs which would not hatch. And it wasn't just lapwings; we saw widgeon, teal, Brent geese and golden plover. Remember, this is all happening within a few yards of the busiest car park in North Norfolk.

There is something about the lapwings that makes them a kind of symbol for a lot of what regenerative farming aims to achieve. And there is something cheerful in the jaunty quiff of erect feathers on the male birds, an exuberant banner proclaiming their readiness to mate. Both males and females sport the same 'hairstyle', although the male crest is usually a little taller. Those wispy plumes join onto a neat black cap on the top of the lapwing's head. A dark beard decorates the bird's 'chin' below its beak, while the throat and sides of the neck are clean and white. Meanwhile, its back is decorated with a shimmering cloak of gloss green and purple, as though the bird is wearing a finely wrought carapace of enamel. The upper breast is neatly bibbed in black, leaving the belly feathers white. Strong, slightly squared-off wingtips are fringed in plain black, and, if you are close enough, you will make out a flash of chestnut on the rump, especially when they dip their heads into the mud to feed. Outside the breeding season it is harder to tell the males and females apart as the colours fade somewhat, although both sexes are handsome creatures. Males and females share the care of the chicks and are famously brave in defending the young from intruders; flocks of lapwings will band together to mob a bird of prey and even scare off cattle that come too close.

Known in different times and places as 'Peewits', 'Green Plovers', 'Hornywinks' and, once upon a time, 'Peesweeps', these birds were once common over much of the countryside, one of the species any rural child would have seen frequently and instantly recognised. In Norfolk they called them 'Pie-wipes'. The distinctive rising cry, to me more guttural and full-throated than a *pee-wit*, is more like the kind of sound made by a squeaky toy, or, when several birds are together on the ground, it sounds like a frenzied game of Space Invaders, the electronic arcade game from the 1980s. Once they were commonly seen in large flocks, but now in many parts of England they are found only in single pairs or in small groups of less than a dozen birds.

The male lapwing has a fantastically acrobatic set of tumbling moves, rolling and diving and creating erratic zigzag moves through the air. Once they have paired off, the birds tend to nest in loose groups. They have a particular way of landing, dropping gracefully through the air with wings held high and onto their feet, which dangle at the end of those long pink-tinged legs. Their nest is a simple scrape in the ground, perhaps lined with a little straw or grass. Usually, there will be four pale khaki eggs covered in irregular dark brown speckles and very hard to spot until you are right on top of them. If disturbed when they have chicks, the adults run away from the nest and only take flight when they are some yards away, making it hard for a predator to pinpoint the exact position of the eggs or the young. They are also one of the species that will feign a broken wing, running along the ground and dragging it beside them rather than flying off at once. The hope is to dupe the fox into thinking that it has an easy victim, and lure it away from the nest site.

Sadly, in recent decades the English countryside has not been friendly to lapwings. They have fared slightly better in Scotland, but even there their numbers have started to decline over the past few years. And, once again, it is almost certain that these birds have suffered because of the intensification of farming.

Once upon a time it was egg collectors who threatened the population, until Parliament passed the Lapwing Act in July 1928, imposing a fine of £5 for anyone killing or selling the birds. Parliamentary protection brought some relief, and a brief upswing in lapwing numbers. At the time they were taken mainly for food, a hangover from the Victorian craze for eating 'plovers' eggs' which had brought about significant declines in the population. Mrs Beeton had several recipes for plovers' eggs, commenting that the white turned 'a beautiful translucent bluish colour' after boiling. In 1953, when the Protection of Birds Bill was being debated, Earl Jowitt remarked that in 1928 he had owned a farm on the Romney Marshes where he would see up to forty lapwing nests. At the time, before it became illegal, he was used to collecting the early eggs so that the birds had time to lay a second clutch when the corn was coming in, when they would have more cover in which to hide. Jowitt said that on the same farm, after the war, it would be hard to find four nests, and he claimed that taking the plover's eggs early was good for the population. In his view, numbers of various bird species had declined because they were being overprotected. If more eggs were taken from their nests, they would breed more strongly. He was put in his place by the noble Viscount Lord Buckmaster: 'The plover knows best when to lay its eggs, and we make a great mistake if we interfere with nature.'

With or without help from the aristocracy, the

lapwings have declined in England by around two-thirds since the 1980s. On the Continent they have not fared well either, with numbers down by half in the Netherlands, Germany and Scandinavia. In these countries, a concerted increase in the drainage of wetlands to create arable land has been a significant factor, while reforestation or the creating of commercial tree plantations has also been bad news. Lapwings avoid nesting near the treeline as they know the forest harbours predators. They recognise the dangers of an 'edge'. We know that they can tolerate nesting on land where animals graze, although heavily stocked areas may be compacted by hooves and this is disastrous because it makes it harder for older chicks to find an essential food – earthworms. When the chicks are very small (in the first two weeks) they like small insects and beetles, which are found in grassland, but as they get older they prefer worms, which are easier to forage in damp earth. The complexities of lapwing survival have made it difficult to devise successful strategies for their protection. Unimproved grassland that has not been treated with pesticides, with a plentiful supply of insects and low predator numbers is one set of factors, but wetlands and damp areas where beetles and insect larvae are within reach of fledging chicks seems to be another. Unfortunately, arable farming hates boggy, damp fields.

A major study of lapwing populations in the Netherlands and Germany showed that protecting nests is not enough to improve population numbers. Simply guarding the nests only increased the population growth rate by two per cent. A bigger improvement in chick survival numbers was brought about through fencing off nesting areas to prevent terrestrial predators visiting them. However, the study points out that fencing off large areas is expensive, and while it may keep out otters or foxes, there may be a downside – avian predators may learn to associate fenced areas with increased numbers of ground-nesting birds and use the 'reserve' as a larder. Raptors have even been shown to learn that when nests are marked by sticks on arable land they are a signpost to an easy meal. The most effective strategy, it seems, is to create areas large enough for the lapwings and other wading species to create a nesting colony. This is the lapwings' natural strategy and it allows the birds to raise the alarm when threats appear. The most successful areas appear to be mixed habitats, with a combination of bare ground, flooded areas and nearby dense vegetation. This means the chicks have both food and shelter.

In North Norfolk we get large flocks of lapwings coming in from elsewhere in continental Europe to overwinter. Our own 'British' birds sometimes stay for the winter, but some move west, over to Ireland, or even south as far as Spain and Portugal. In the winter of 2020–2021 we counted record numbers of lapwing around Holkham, with more than 17,000 here for the first time in many, many years. It seems very much to be a clear case of 'build it, and they will come'.

9
Brick O'Longs

The declining numbers of birds in our countryside has not gone unnoticed. In 1962, the British Trust for Ornithology set in motion something called the Common Bird Census, a scheme that aimed to count certain species at a list of regular sites. It provides cumulative data on some 164 species, and it was the first national monitoring scheme for breeding birds in the world, running until 2000. The scheme merged with the Breeding Bird Survey, incorporating the data collected by an even larger team of volunteers who keep an eye on over three thousand measured sites between April and June each year. Another record, the Farmland Bird Indicator, takes those results and looks at individual trends for almost thirty species which might be expected to be 'found feeding in open farmland during the breeding season'. It was decided to exclude winter bird numbers, as many migrants arrive to shelter here in sizeable flocks (like the lapwings) but do not stay to breed. These overwintering bird numbers fluctuate according to how well they have done in their 'home' territory. From this list, Defra has, in turn, chosen nineteen species to use as an average annual index.

The latest figures for trends between 1970 and 2019 are not encouraging: farmland birds are down by 57 per cent; woodland and seabird populations have both declined by around 27 per cent. The BTO puts considerable effort into making these surveys authoritative, and Defra applies statistical science to make sure that the trends they record have a reliability, or 'confidence interval', of 95 per cent. They use what mathematicians call 'smoothed trends' to reduce short-term peaks and troughs in numbers which might be due to individual year-to-year weather variations, and differences in breeding success due to unusual or one-off circumstances. Defra's figures show that in the past few years (2013 to 2019) out of some 130 species surveyed, 32 per cent of species showed an increase, while 37 per cent showed a decline.

The most dramatic declines have occurred since the 1970s, and the most recent numbers show a slower rate of loss. It's been a good period for Cetti's warblers, great spotted woodpeckers, red kites and collared doves. It's been a very bad time to be a turtle dove, a grey partridge, a tree sparrow, a lesser whitethroat or a willow tit. This last group are now all at less than a tenth of their 1970 strength. Worryingly, while seabirds, woodland birds and water and wetland birds have all shown gradual falls in their populations, it is the farmland species which have declined fastest and in the largest numbers. Given, as I have said before, that more than 70 per cent of our country is farmed, it isn't hard to put two and two together, is it? Something is going wrong on the farm.

The biggest declines were between 1970 and 1985; since then the numbers have continued to fall, albeit more gradually. But before we persuade ourselves that this is a good sign, it's worth recognising that for the farmland birds 21 per cent of species have shown a 'weak increase' since 1970, 16 per cent have remained stable, but 63 per cent have 'shown either a weak or a strong decline'.

The data from 'the men in the ministry' provides very straightforward clues as to what is happening. They talk about 'known and potential causes' and point unequivocally to the decline in mixed farming, the move from spring to autumn sowing of arable crops, the switch from hay to silage production on grassland, increased pesticides and fertilisers and, my obsession, the removal of hedgerows. It all adds up to less food, less suitable areas for nesting, and not as many feeding opportunities. Defra itself states that some farmers are doing what they can to reverse these trends by providing overwinter stubble, planting wild bird crop covers and 'sympathetically managing' their hedges. But hidden within the relatively hopeful news is a stark reality: some farmland birds are classified as 'specialists', meaning they are restricted to, or are highly dependent on, farmland habitats. Think of the grey partridges and their reliance on field margins, arable seeds and being able to hide their chicks in the right sort of crop. These are the species showing around 70 per cent declines. Defra has singled out five of these specialists – grey partridges, corn buntings, starlings, turtle doves and tree sparrows – all of which have declined by more than 80 per cent since 1970. I find that pretty shocking.

The full list of farmland birds on the ministry list is as follows:

Tree sparrow (*Passer montanus*)
Corn bunting (*Emberiza calandra*)
Turtle dove (*Streptopelia turtur*)
Grey partridge (*Perdix perdix*)
Yellow wagtail (*Motacilla flava*)
Starling (*Sturnus vulgaris*)
Linnet (*Linaria cannabina*)
Lapwing (*Vanella vanellus*)
Yellowhammer (*Emberiza citrinella*)
Skylark (*Alauda arvensis*)
Kestrel (*Falco tinnunculus*)
Reed bunting (*Emberiza schoeniclus*)
Whitethroat (*Curruca communis*)
Greenfinch (*Chloris chloris*)
Rook (*Corvus frunilegus*)
Stock dove (*Columba oenus*)

These national statistics are all worrying, but at the local level it is possible to bring about improvements and, I maintain, it can be done quickly. I see many of these birds at Great Farm. By taking the unproductive fields out of arable production, planting bird-friendly wildflower seeds and allowing the hedgerows to do what they do naturally, we have already begun to see change.

As well as the relatively small area which has been taken out of production, the overall condition of the soil is being improved by a range of cover crops, including mixtures of the following species:

Species	Purpose
Sainfoin (*Onobrychis*)	Drought-resistant, needs no nitrogen fertiliser, does not cause bloat and is a natural anthelmintic (kills parasitic worms). Frost hardy and suits thin soils.
Winter vetch (*Vicia villosa*)	Winter hardy, produces large amounts of nitrogen. Excellent rooting crop.
Red clover (*Trifolium pratense*)	Relatively winter hardy, rapid growth, strong tap root, produces nitrogen.
Bird's-foot trefoil (*Lotus corniculatus*)	Natural anthelmintic, used mainly for livestock benefits as it can also improve protein utilisation.
Alsike clover (*Trifolium hybridum*)	Later flowering than other clovers, good nitrogen fixer.
Crimson clover (*Trifolium incarnatum*)	Annual clover but can reseed itself. Fast growing.
Lucerne, inoculated (*Medicago sativa*)	Excellent rooting, even on drought-prone soils (soil needs to be alkaline / free-draining). Better in second and third years when the forage yield is maximised.
Sheeps burnet (*Sanguisorba minor*)	Early growth in spring, very deep rooting and stays palatable for a long time on thin, dry soils. Roots will bring up nutrient from depth. Larger seed than some so seed rate may look high (compared to sheeps parsley, for example).

Sheeps parsley (*Petroselinum sativum*)	Deep-rooting with a strong tap root, drought tolerant and good for soil conditioning.
Ribwort plantain (*Plantago lanceolata*)	A mineral-rich plant that draws up nutrient from depth via strong roots. Very drought tolerant.
Chicory (*Cichorium intybus*)	Strong, penetrating root that can reach great depths and break through compacted layers. Enjoys dry soils. Natural anthelmintic.
Yarrow (*Achillea millefolium*)	Small seed but the plant produces a strong root, very compatible with sheeps parsley, burnet and plantain. Another very drought-tolerant species.

These mixtures will be planted following the harvesting of spring barley in fields scheduled for spring cropping the following year. It's part of the plan to make sure that there is never bare soil left on the farm and that there are always living roots in the fields. It's well established that a multi-species cover crop combined with a small amount of grass does better than a monoculture as the plants are strengthened by some competition. A recent study in Cambridgeshire found that maize and wheat fields sown with cover crops had three times as many earthworms afterwards. The worms, which in themselves are an indicator of soil quality, also did better, and there were more of them when the soil was disturbed as little as possible by tilling or cultivation.

The main experiment, or demonstration area for what I'm trying to do at Great Farm, is the field at the north-west corner, marked on the maps as Brick O'Longs. At ten hectares (25 acres) it's about 10 per cent of the total farm. It's got eight corners, and is a nightmare to drive around with any kind of agricultural machinery. It slopes, there's a big dip in the middle to the northern side where there is what's known as a SHINE (Selected Heritage Inventory for Natural England) site. That means it's registered on a database of English archaeological sites which are classified as suitable for one of the numerous government-funded programmes that pay landowners for conservation, known as Environmental Stewardship. At Brick O'Longs we don't know exactly what's underground at the SHINE site, but it needs to be protected and preserved as part of the historic environmental assets. The hollow and its mysterious contents are covered with dense scrub and there is pooling water underneath it all. The irregular corners, the pit and the clay soil – probably the heaviest on the farm – combined with the quite severe slope from west to south-east, and a seasonal underground spring on the eastern edge, make it quite uneconomic for farming. Harvesting of sugar beet by the previous tenant had compacted the soil and, when heavy rains came one March, a large quantity of topsoil slipped into an adjacent property, a modern home built from old converted agricultural buildings.

As part of the plan to restore nature to a typical farm, we devised a systematic plan of action, including baseline surveys of what we had on the land. This included fixed-point photography, daily weather records and management of predatory species. We carried out night-time deer and hare counts in February and March, as well as winter bird surveys using BTO methods. February also saw us participating in the big farmland bird count organised by the Game and Wildlife Conservation Trust. In April, we carried out small mammal trapping for identification and population estimates. April and May saw five separate farmland breeding bird counts and a butterfly transect survey. May and August were when we carried out bee and botanical counts, with June and August seeing moth trapping. September was a busy time, with a fortnightly butterfly transect, hedgerow assessments, as well as woodland understorey surveys. October always brings a daily hare count (when we also count any other mammals seen).

After the field was cleared of commercial crop and had dried out enough, it was cultivated and sown with a seed mix containing mustard, fodder radish, white millet, phacelia, quinoa, red millet, buckwheat and reed millet. This mixture would cover the degraded soil and protect it over the summer. It would also provide a significant range of winter bird food. Obviously, the food dwindled as winter wore on, but there was always something left for a passing visitor, as it covered a reasonably large area. By the time the seeds were finished and new growth was coming in on the farm, we had applied for a Stewardship Scheme, which allowed us to cultivate the 'annual 'weeds', the plants that create a particularly attractive structure on the land which is vital for ground-nesting birds like stone curlew, lapwings, oyster catchers and skylarks. My approach was to make sure there was always something in the feed for the birds, and even after cultivation we left some areas – about four hectares – as hay meadow so there was still seed for the birds all through the summer. One of the things I want to demonstrate is that regenerative – or, perhaps more accurately, *restorative* – farming and the provision of food for birds and invertebrates, is not the same as rewilding, or leaving nature to do its own thing. It's also not necessarily a process that follows organic principles in terms of making sure no artificial inputs are ever put onto the land.

Sometimes, removing land from production is as simple as squaring off the planted area to make the use of agricultural machinery more efficient. For example, at Tinkers Hole, almost one hectare in that single field was made available for hay meadow because the driver of the crop sprayer was allowed to devise the most efficient (squared-off) shape on which his machine would operate. The boom of the modern crop sprayer extends 18 metres either side of the cab, allowing a 36-metre spread in total. He can accomplish his task much more rapidly if he drives it in straight lines and doesn't have to manoeuvre it back and forth in awkward corners. The cost of operating this high-tech machinery makes it much more economic to use it as efficiently as possible. On most farms there will be land which never produces as much crop as others, whether it's because it is shaded by a line of trees, or has a boggy hollow somewhere which is difficult to harvest. Making land available for nature doesn't mean squandering productive areas. In some of the media coverage of the government's plans for Environmental Land Management, there has been a degree of fearmongering about 'abandoning farmland in favour of rewilding'. Stories about the dangers of 'a fad for rewilding' often cite examples of wealthy individuals who wish to plant trees on their land, and equate such projects with an inability to grow the food we need. I'm advocating something which is very much integrated into our food production process.

At Brick O'Longs, the field was divided up into three strips, running south to north, lengthways. On the western side, the growth was cut and cultivated – the process where some kind of mowing machine removes the aerial part of the plants and stops seed dispersal onto the field. In the central portion of the field (i.e. the middle strip of the three not turned over to bird seed), the plants were topped, then sprayed with glyphosate and then cultivated – lightly tilled. The remaining eastern strip was just topped and not cultivated, and no glyphosate was added. Glyphosate is a contact herbicide labelled by chemists as '*N*-(phosphonomethyl)glycine', but is known to most gardeners as the main ingredient in the weedkiller marketed as Roundup®. It now goes under many different trade names, and the manufacturers state that it's absorbed primarily through leaves, or the parts of the plant matter it touches, and only minimally reaches the roots. Within soil and water it seems to dissipate relatively quickly.

Some agricultural experts see glyphosate as one of the most important discoveries in farming science and compare it to the arrival of penicillin in human medicine. Others claim it is potentially toxic and carcinogenic to human beings, although it works by interrupting chemical pathways within the plants which are dissimilar to those in the human body. The FAO and the WHO, as well as many other regulatory bodies, have said they do not believe it poses significant risk to human health, although some countries have now banned its use, or soon will. Once glyphosate is applied to plants, they stop growing within a few hours, and over a few days the foliage turns yellow. Farmers often use it as a desiccant to dry down the foliage before they harvest the root crop underneath, which makes the process easier and more efficient.

The application of glyphosate saves farmers the necessity of deep ploughing, which in some soils is harmful to the most fertile layers of soil and may result in greater moisture loss. Without a desiccant like glyphosate, more ploughing would probably be necessary to control weeds. After a species-rich cover crop has been planted, for example over winter, the farmer wants to remove it to plant the new season's crops without excess disturbance of the soil. Spraying the cover crop with glyphosate is the easiest, most efficient way to do that. As glyphosate doesn't affect the seeds, which have not yet germinated, the new crop can be drilled into the soil either before or after spraying. Meanwhile, the residue of the cover crop will protect the emerging food plants from extreme heat, cold and rainfall, and also give them a shield against pests while they are very small. Even while the leaves wilt and die after being sprayed, the roots of that cover crop remain undisturbed in the soil and are storing carbon. Those roots are also improving the structure of the soil, while retaining moisture. In the context of regenerative farming, if I have a field margin which I just lightly cultivate, in preparation for the restoration of annuals, gradually the grasses will begin to dominate and outcompete everything else. But glyphosate allows me to kill off the grasses and let other species come through.

At Brick O'Longs, the westernmost strip produced a great diversity of vegetation. It became a vibrant carpet of poppies, cornflowers, speedwell and mayweed. These are species that complete their life cycles over the course of one calendar year. Many, if not most of them, are archaeophytes, plants which were not original to Britain (or Ireland). Many of them came during the Roman occupation, though some were brought here as long ago as Neolithic times. Botanists distinguish between the archaeophytes – which were often food plants – and 'modern introductions' which were often ornamental plants or species of timber brought over during the ages of exploration of the New World. The archaeophytes tend to favour man-made habitats like agricultural land rather than native forest or so-called wilderness. Introduced or not, we regard many, many of these plants as emblematic of what we believe to be natural in the British landscape. Just as we now fret about the demise of the cornflower or the scarcity of fumitory, some of the plants familiar to previous generations of country dwellers have also all but disappeared from wild settings, among them the downy hemp-nettle (*Galeopsis segetum*) and the stinking hawk's beard (*Crepis foetida*). Some, like the lamb's succory (*Arnoseris minima*), are probably extinct in the arable landscape. Others, like the corn-cockle (*Agrostemma githago*), are country casualties of the practice of growing winter wheat, cut down before they could seed and ruthlessly eradicated from the corn by more efficient cleaning techniques during the harvest.

These archaic yet familiar species tend to grow well in land which has been lightly cultivated, allowing the penetration of moisture and light. They also prefer it when it's not too richly fertilised, not necessarily because they can't tolerate nitrogen-rich soil, but because more fertile soil allows more aggressive species to outcompete them.

The fact that these plants have thrived at Great Farm is a sign that what we are doing with the land is effectively mimicking nature to some extent. These were the plants which were already in the soil seed bank, and began emerging in the spring as the soil gradually warmed. No herbicides had been sprayed, and young leverets were now always to be seen feeding on the young plants. In high summer, this is the richest area of the field in biodiversity terms, and full of the sounds of insects. I often find roe deer lying in the thickest vegetation, raising a head to peer out when they hear me approaching.

In summer, the least variety of life was found in the middle strip – which had been topped, sprayed and cultivated. However, in spite of delivering very little in spring and summer, it still had an abundance of plants that would provide seeds over winter. The plants that sprang up were the residue of those that had been planted the previous year – the millet and fodder radish and phacelia – all of the things that provided a pollen source in summer and then seeds in late summer and through winter for the linnets and yellowhammers. There were ox-eye daisies, viper's bugloss and red clover, too, which had spread into the area from previous grass margins. Walking through this area it was very clear, even to a casual observer, that it was full of bees, ladybirds, butterflies, aphids, hoverflies and pollen beetles. On Great Farm we have already had butterfly surveys which have found seventeen UK species – small skipper, large white, green-veined white, small white, small copper, brown argus, common blue, red admiral, painted lady, small tortoiseshell, peacock, speckled wood, wall brown, gatekeeper, meadow brown, small heath and ringlet.

In the summer of 2021, we also carried out a detailed bee survey on Great Farm, principally at Brick O'Longs. Forty-five species of bees and wasps were found, many of them without common names. Notable were the presence of *Colletes marginatus*, a solitary bee which feeds only on hare's-foot clover and yellow and white clovers. There were cuckoo bumblebees (*Bombus vestalis*), so called because they take over the nests of the much more common buff-tailed bumblebee (*Bombus terrestris*) and assassinate the queen. It, too, needs clover and vetch, blackthorn and dead-nettles, all the kinds of plants we have around Great Farm, and especially at Brick O'Longs. Several of these insects are rare in the UK, and some are especially rare in Norfolk. Some, including the buff-tailed bumblebee, are relatively common species, nevertheless they are increasingly abundant and seem to remain active even in winter. These classically obvious bees have vivid yellow bands across the middle of the abdomen and at the front of the thorax. Others are relatively new arrivals to Britain, species like the tree bumblebee (*Bombus hypnorum*), which is common in continental Europe, or even further afield, and has become one of the most prolific species in England and Wales. Recognisable by the rich ginger hairs at the tip of the thorax, like an elegant fox-fur stole, it also has a white 'tail'. This provides more evidence of the constant flux we witness in the natural world, and the changing climactic conditions in the south of England that are allowing these once exotic visitors to thrive. Great Farm has been found to support the clover blunthorn bee (*Melitta leporina*), which is strongly attracted to white clover and likes to roost on the underside of flower heads of the commonly seen yarrow (*Achillea millefolium*). The list goes on, with sharp-tailed bees (*Coelioxys conoidea*), which to most people look more like wasps and are another parasitic species that lays its eggs in the nests of mason bees (*Osmia bicornis*) and some other species. They collect no pollen, leaving their larvae to feed on the pollen collected by their hosts, and after they have killed off the mason bees' own larvae.

They love the perennials, knapweeds, thistles and

ragwort. These, of course, are the species of plant that make farmers nervous, and the ones which have traditionally been the target of eradication and can, in some cases – such as ragwort – be toxic to animals. A small number are classified as 'injurious or noxious' weeds' under the Weeds Act (1959), including common ragwort (*Senecio jacobaea*), broad-leaved dock (*Rumex obtusifolius*) and curled dock (*Rumex crispus*), as well as the creeping and Scotch thistles (*Cirsium arvense* and *C. vulgare*). It's not illegal to have them growing on the land, and Defra acknowledges that they have significant environmental benefits, but farmers are duty-bound to stop them spreading onto neighbouring farms. These species can, however, be managed agriculturally, as we do at a site we call the Burrows, where they occur in small numbers among the other arable plants but do not dominate or outcompete the other species.

Great Farm hosts many species of wasps, too, many of them without common names, like *Diodontus luperus*, which loves to eat aphids. Meanwhile, *Cleptes nitidulus* preys on sawflies, and the rare heath wasp, *Cerceris ruficornis*, eats weevils, although its life cycle is unknown. In other words, there is a lot going on here with the invertebrates, and plenty of them are beneficial species which eat pests, and all of them of course are potential food for the birds too. What we're seeing is species that we barely think about and which only experts can identify accurately. But you don't need to be a wasp specialist to notice the sheer volume of insects attracted by the wide range of plants at Great Farm. Nearby, there are other farms where the fields resemble deserts – tracts of mud in winter or virtual dustbowls in summer, where great cracks appear in the soil as it dries. These fields support no birdlife because there are no plants to provide seeds or to attract insects.

Great Farm is turning into a hot spot. What I want to see is a network of areas like it criss-crossing the countryside, coexisting with working farmland where food crops also grow. There is, aside from the biodiversity benefits of having fields like this, another important and all too often overlooked human aspect. The man who operates the combine harvester on Great Farm told me in late summer that he looked forward to working the land here because he said there was always something to look at. In the industrialised, intensively farmed landscapes which we have created, we forget that for many farmworkers the connection with nature has been all but removed. They are not seeing the birds, the hares, the deer. They might as well be driving a taxi in town for all the connection they get with nature. It's been said countless times that people will only protect what they love, and they can only love what they know. Along with those of us who seek to visit the countryside for pleasure, those who work on the land deserve a better natural environment too.

Even though we are in the early days of the experiment at Great Farm, the proliferation of flowers, insects and other invertebrates is already bringing a change in the avian populations.

Great Farm	**2020**	**2021**
Blackbird	0	2
Blackcap	2	2
Blue tit	0	2
Chaffinch	2	7
Chiffchaff	1	2
Coal tit	1	0
Dunnock	4	8
Great tit	1	0
Goldfinch	0	2
Grey partridge	2	1
Lesser whitethroat	2	4
Linnet	0	4
Long-tailed tit	0	2
Mistle thrush	0	1
Oystercatcher	0	1
Quail		1
Robin	1	2
Skylark	16	22
Whitethroat	0	9
Yellowhammer	5	13
Totals	**37**	**85**

The numbers in this chart are not individual birds – but *breeding pairs* of these named farmland species counted on a single day. Even within the space of this short period of new management principles at Great Farm the total number of breeding birds spotted in the same fields and hedgerows has more than doubled. Nesting whitethroat, long-tailed tit and oystercatchers had simply not been seen on this farm before. The yellowhammers have been especially welcome, and their resident nesting population has almost tripled. Birds which rely on seeds and berries, like the goldfinch and the linnet, have also appeared. Yes, the numbers are relatively small, but the trend is almost entirely positive. Remember, this is a farm comprising just ten fields and little more than 100 hectares, which had been very intensively worked and had suffered the application of many different sprays and treatments to remove unwanted life. It had no new margins, no cultivated headlands, no floristic margins and no wild bird cover. The changes we implemented were designed to make the whole farm more biodiverse, and the results exceeded our hopes, especially in terms of how quickly we began to see positive change.

It wasn't just the birds that came back to Great Farm. By August 2020, an insect survey revealed more than a thousand individual butterflies from seventeen UK species were present at Holkham over a twelve-week period. August was the most prolific month, with a very large cohort of small whites (*Pieris rapae*), an amazing sight for me, their pale wings decorated with a small brown dot and the leading edges smudged with the same colour, as though they had been dipped in demerara sugar. A survey of bees at Great Farm by the Norfolk Naturalists group found that there were eleven of the UK's twenty-four species of bumblebees present, including the largest of them all, *Bombus ruderatus*, the large bumblebee. This has a very long tongue, adapted to feed from tubular flowers. The species has declined sharply in recent years and is now classified as a Species of Principal Importance under the Biodiversity Framework (previously the UK Biodiversity Action Plan).

Like many of our farmland birds, the large bumblebee has been affected by intensification in agriculture and urban development, which together have led to the loss of the buble-bee's preferred habitats: flower-rich grassland, wet grassland and ditches. It especially thrives on unimproved flower-rich areas, places in our efficient, developed landscape which are increasingly fragmented and small in area. Its favourite foods include comfrey, teasel, red clover, toadflax, dead nettles and thistles. This bee has gone from 80 per cent of its former range, partly because the queen bees need to find the abandoned burrows of small mammals in which to hibernate. Those nesting burrows also need to be within reasonable foraging range around 500 metres – of the pollen and nectar supplies the bees need.

The breeding birds, the butterflies and bees, the hedgerows and wildflower strips and the timing of farming activity all play their roles in making the farm more nature friendly. It's more evidence that a whole-farm approach to creating biodiversity is better than just leaving a few patches for nature at field edges or in the odd unused corner. Along with that whole-farm approach, we need to collaborate so that neighbouring farms cooperate in doing the same things with the same nature-friendly goals. That's where the massive restorative gains for nature will come.

10

The Long View

The Burrows
August

At Warham Camp, the spiders have woven funnel-shaped webs in the grass, like fairy goblets spun from gossamer left behind after a party. Labyrinth spiders lurk in the dark depths of these narrowing tunnels where they hatch their young, waiting to pounce whenever a small insect, preferably a grasshopper, makes the mistake of entering the tube. Tickling the entrance to the web with a blade of grass brings the little spider out to investigate, revealing its delicately speckled caramel-brown legs and symmetrical chevron body pattern. They're harmless, but sometimes people find the web and immediately imagine they must be similar to the deadly funnel-web spiders of Australia. There are thousands of these tiny labyrinth spiders here on the chalky doughnut rings that are all that's left of an Iron Age fort. The concentric earthwork rings are incomplete, with much of the western edge missing and possibly never as high as the rest of the circular embankments. On that side, the Stiffkey river itself would have been enough of a barrier to any enemy. Some say the wall did once extend in this direction but the soil was dug out and used to create a bund to divert a short section of the river below. Around the banks of the river are fields which have been industrially farmed for many years, planted with wheat and barley, sugar beet and potatoes.

High above the waterline, for a brief time each summer the site is a hatching ground for thousands of delicate chalkhill blue butterflies, which come to feed on the horseshoe vetch that grows on this poor soil. They flutter here and there on pale, blue-rinsed wings that fade like a spray-painted decal into an earthy brown rim, outlined with a neatly edged white border. The grass is peppered with the tiny flower heads of squinancywort, used in medieval times as a cure for sore throats (when it was called quinsywort). Cattle graze here, though they favour the lea field edges closer to the river where the grass is lush.

It strikes me very strongly that this river valley is a landscape which is constantly evolving. Once upon a time, these fields were regularly flooded, an example of a rare landscape feature called an *alder carr*. Still marked on the Ordnance Survey maps as 'the Carr', there's a scrap of this habitat visible just to the north of the narrow channel, which is all that can be seen of the River Stiffkey as it veers eastwards before flowing towards the coast. It's a waterlogged and wooded terrain which has become vanishingly rare in England, thanks to modern methods of land drainage. Shrubs rather than mature trees are its distinguishing feature initially, but gradually it accommodates trees like willows and alders, which can tolerate water. The English word 'carr' stems from the Norse *kjarr*, meaning something close to scrub. The settlers from the north-east have left other marks on the land here.

Once this place was called the 'Danish Camp', and at other times Barrow Hill Camp. Now, it's referred to either as 'the Burroughs' or 'the Burrows', and which is more correct has been lost in the mists of time. There is a theory that it gets its name from when a warrener had rights to the area, because it was rich in rabbits which colonised the raised dry ground. This fort has a diameter of just over 200 metres (a little smaller than the one next to the spoonbill colony at Holkham), but it's still on estate land, just further inland and near to the village of Wighton a few miles south of the coast. The circular embankment consists of double ramparts which today are about four metres high, enclosing around three acres. It's a peaceful spot, sitting in a crook of the Stiffkey, one of England's 200-odd chalk streams, and one of only four in Norfolk which flow into the North Sea. Once upon a time, these spring-fed streams bubbling up from the chalk aquifer were famed for their clear water and unique wildlife, and valued as important breeding grounds for both brown and sea trout.

There are thought to be fewer than three hundred chalk streams in the world, more than 85 per cent of them in England, stretching in a diagonal band roughly west to east from Wessex to Norfolk. There are also some further north in Lincolnshire and Yorkshire, and a handful in France around the Somme. The Stiffkey, like so many of these important habitats, has suffered from sewage inflows, a weakening stream caused by water being taken for agricultural use, and general pollution, especially run-off from agricultural land that has been over-enriched with nitrates, phosphates and other fertilisers. All of these things, including the diversion of many rivers as far back as medieval times, have led to environmental degradation and the loss of once crucial habitat for otters, water voles, crayfish and mayflies, as well as the plants that grow in them, like starwort and marsh marigold. Just a small number of the remaining English chalk streams are protected.

To the west of the river arm, the land was once called Sweyn's (sometimes Svein's) Meadow, giving another clue to the Viking settlements here. It seems Saxons from nearby Wighton used to periodically poison the river here to drive the Danes away. In truth, the Burrows was a fort long before the Vikings arrived in Norfolk towards the end of the ninth century, and excavations in the summer of 1914, in the 1920s and again after the Second World War, found several pieces of first-century Roman pottery, a brooch and some even earlier fragments of crockery. Other artefacts, collected haphazardly in the early twentieth century, were in the possession of George Youngman, a local marshman and molecatcher who, naturally, had done a bit of digging in the area. He also said that human bones were periodically found here. Once again, we see how important evidence comes to us from the eyes of a man who is out and about on the land.

To the north of the fort there is a ten-hectare field that we also call, simply, 'the Burrows'. It's a piece of sloping chalk grassland, and for many years it was used as a cultivated arable field, mainly growing barley. Historically, the location and chalky substrate meant it was regarded as poor land for grazing. In winter, a few cattle or sheep might have been put there to benefit from the well-drained soil which, unlike lower-lying ground, wouldn't get churned up or poached by their hooves in the wetter months. Notwithstanding, it's relatively poor soil, but less than a decade ago this was a cultivated field. Like parts of Great Farm, it's simply not justifiable in economic terms to put the effort into farming this type of land.

In the adjoining field, behind the hedgeline, the area around the fort is officially a Site of Special Scientific Interest and, long before such things were classified, the hummocks of the fort made it unsuitable for ploughing. About eight years ago, as part of a subsidised nature scheme to improve calcareous grassland habitats, the Burrows field was emptied of livestock. A team of plant specialists collected seeds laboriously by hand from the local area (no further than twelve miles away) and, of course, from the field next door containing the fort. Holkham Estate sowed the seeds and the field was managed simply, by mowing and then removing the cuttings. No chemicals or fertiliser were put onto it. However, when I came to Holkham I discovered that questions were being asked as to whether it was economical to keep the field in this improvement scheme. I suggested that instead of spending money cutting the field, it would be cheaper, and potentially better for its floral diversity, if we grazed it with a small number of sheep which would all be removed by April. Some old hands in the naturalist community were horrified, accusing me of ruining all of their good work. They suggested the field should be topped and left to gradually return to scrub. But this is not a nature reserve, it's working farmland, and it seemed more sensible to see if livestock could graze these fields and yet still allow them to bloom with a huge variety of meadow plants. This way, the sheep get fed, there's no manpower or fuel used in cutting the meadow in August, and the animals create a more varied structure in the existing vegetation. In my view, it's not always necessary to go the pure rewilding route to help nature, and while it works in some places, it's not always the best solution for wildflower meadows, which seem to thrive when there is a livestock component to their annual cycle. In the Burrows, I took some botanists to visit the meadow in the first spring after that winter of light grazing by the sheep. Whereas in the previous year there had been slightly fewer than three hundred pyramidal orchids, this time we counted more than a thousand. There were also healthier numbers of common spotted orchids and bee orchids. To me, this was evidence that sometimes things can be achieved by being a little more flexible than strict science – or farming wisdom, for that matter – dictates. Ecologists classify hay meadows and permanent pasture as 'early succession habitats' and calculate that they can support almost eight hundred species of wildflowers and around twice as many species of pollinators. Leaving land to revert to nature, however, is more likely to allow it to turn into scrub and, eventually, woodland. The natural disturbance of grazing allows the wildflowers to emerge and not be drowned by the taller, woodier growth that would soon outcompete them.

Last summer, a walk through the Burrows revealed a mass of flowers and insects, a profusion of life that rivalled, if not bettered, the SSSI next door. By July it was filled with tall fluffy yellow flower stalks of lady's bedstraw with their distinctively hay-like scent. The wild grasses were abundant – wispy tufts of yellow oat grass swaying in the wind alongside sweet vernal grass, a favourite in the old days not just for cattle, but of farmers who would chew it for its mild vanilla-like flavour. It's less favoured by hay-fever sufferers, but it's an important food for the larvae of brown and skipper butterflies. There was quaking grass, called 'dithery dock' or 'tottering grass' in some parts of the country because its flowers are carried on heart-shaped spikelets which flutter in the slightest breath of wind. Walking through the grass was to be tickled by the thick growth, and picking out the litany of different species was dizzying . . . Queen Anne's lace (wild carrot), wild parsnip and tall, bright yellow agrimony, a traditional country remedy for inflammation. The violently pink heads of another butterfly favourite, black knapweed, stood out like flashing beacons among the greens and yellows.

There is poetry in the common names of so many of these plants. The Burrows has plenty of crested dog's-tail, a stiff-stemmed grass with staggered flower spikes that carry a delicate purple blush at the tips. Any time before the war people would have known it as a favourite, planted deliberately for drying, bleaching and soaking so that in winter the plaited stems could be used to make hats and bonnets. Its emerging lower leaves were very appealing to sheep, which left the tall flowering stalks alone, and it is also a food for brown and skipper butterflies. I see them often in this field in July and August, which is also the time when the flower heads and grass stems are covered in six-spot burnet moths. These little day-flying insects are about an inch long, but they're more exotically patterned than many butterflies, with their smoke-dark wings decorated with irregular-shaped pairs of vivid crimson splotches. There is something startling enough in that colour combination to human eyes, but it's also a clear signal to predators that they will exude poison if they are eaten. They aren't rare in England at all, but they appear in their thousands here, and within a very short time of emerging from their cocoons they flit from stalk to stalk among the grasses, pausing to mate, tail to tail. They seem to love the bird's foot trefoil, its bright flowers shaped like tiny slippers. Vivid yellow when they open, they have a dark reddy-brown blush at the tips of the buds, which gets them their other nickname – 'eggs and bacon'. Only after the flowers have gone do the splayed seed pods resemble 'bird's feet'. Again, it's not a rare thing; but the connections between the flowers, the seeds, the caterpillars and the moths and butterflies all seems so vivid when I walk through this field which in May, June and July has so much going on all at once. There are only a handful of fields like this left in Norfolk, perhaps 330 hectares in total. Imagine how many insects, flowers, birds and pollinators would survive if that number could be increased.

The Decoy
Late September

The arctic birds have come early this year, driven south by snow. In England, the end of September is still warm and the leaves are barely starting to turn. Dawn is gentle and, at just after 6 a.m., the sky has taken on a faint blush of rosé. The marshes lie blue-grey, a blurry line of silhouettes as if in a watercolour painting. The movement comes from the pink-footed geese, and today there are several thousand taking to the air, ready to group together and set off for a day's foraging. They will come back here to roost in the late afternoon, providing another spectacle for anyone lucky enough to see it. There is a gathering storm of wingbeats and the sharp staccato call of the geese now as they reach a critical mass against the lightening sky. *Cree-ake, cree-ake, creeeake* – the beat is rapid, and takes over from everything else that is happening in front of me. It's like white noise: mesmerising, hypnotic and maddeningly repetitive, so that the birds sound like some kind of giant infernal machine. As the geese rise from the ground they create another dark line, indistinct at first, making it seem as if the far trees are moving and then disintegrating, a pixelating shape as they rise higher and chinks of sky break up the solid shadow. Then the individual wingbeats start to slow, and higher and higher the birds rise, forming at some unknown signal into skeins and arrows, shifting clouds that remind me of that children's toy where you shake the iron filings against a magnetic background, never knowing quite what shape they will form. I know I am lucky to witness this: a mass gathering as spellbinding as watching the wildebeest crossing the Mara River, or a super-pod of dolphins at sea.

There is movement in the world that I can clearly feel. It's not just in the angle of the morning sun, but in the changing smell of the early morning air. Suddenly, with no definitive warning you detect a shift, something in the balance of warmth and freshness, something released from the earth during the night. The movement can come at any time in September, occasionally even in the first few days of October. It may warm again, just a few degrees, but the real summer is ended. The good thing about this time of year is that my body clock means I am adjusting to dawn coming just a little bit later each morning. After more than thirty years of rising half an hour before first light, I appreciate getting up not quite so early as winter approaches, those magic hours in the half-shadow when the change of axis is most tangible. It is a time for contemplation, and for clarifying my plans. Nature brings me fresh ideas.

I am sitting near to the old Decoy Wood, an area of flooded pools where the creek forms an elbow thick with oaks and willows in the middle of the grazing marshes. We're planting more trees to help replace the older specimens, some of which are suffering because of the sheer amount of acidic guano that's built up below the nests of seabirds that roost here. The guano kills vegetation beneath the trees, and also their leaves until eventually the branches deprived of foliage die too. A creek runs roughly east to west and, as well as its watery boundaries, the thicket of trees makes this a protected spot. Come spring and it's the perfect nesting site for colonies of breeding cormorants, grey herons and spoonbills. The spoonbills, with their pure white plumage and elegant spatula-like beaks, have become part of the furniture here, but until 2010 when a small group successfully fledged chicks at Holkham, they had been absent from England as a breeding bird for well over three hundred years.

It's not just the natural environment that seems very present in this quiet spot. There is something palpable in the history here, in the ground where the shadow of the Iron Age fort lies on the land like a scar, and in the marshy ponds where the ghostly silhouettes of the dead and dying trees make perches in spring and summer for the chattering birds. Estate records from 1634 describe my viewpoint as 'a spetiall peece of ground [which] at every high tide and springe tide is compassed with the sea but never overflown'. In the eighteenth century, the decoy was used to trap ducks, and sufficient numbers were caught to support a trade with nearby market towns, as well as supplying the estate with whatever it needed. It's now become a place where life is created, not taken.

I am on the raised ground to the east of the Decoy, a vantage point created by the ruins of the fort, one of several known locally as a 'Danish Camp' and extending almost 400 metres from north to south and 250 east to west. Aerial photographs reveal two small circular depressions within the fort, known as 'pit dwellings'. The Viking threat hereabouts is long gone, but last year, not far from this spot, we had to call in the army bomb disposal unit to remove an unexploded Second World War shell in the middle of where the cows were grazing. More than 2,000 years separate the men who built this fort and those who dropped the bomb. Dramatic as that was, it is farming, not war, that has made even more indelible marks all over this coastline, even driving back the sea and creating pastures from saltmarsh riven with deep muddy channels.

It strikes me that this is a place where change is constant. My plans, which I hope will bring yet more life to the marshes and make the farms a better place for growing food and hosting nature, are just the latest in a long list of human interventions over the centuries. We call it 'Nature' as if it is something other, something with a spirit of its own that exists in another dimension. It's easy to abdicate responsibility and assume that things will somehow muddle through, that the forces of life will adapt and move somewhere else when we alter the land. And yet, we can see the dreadful consequences of many of the things we have done to it when we bend it to our own short-term needs. We may not control that spirit, call it a life force perhaps, but we can work with it to fix some of the damage we have caused in the name of progress. The power of regeneration and the ability of nature to fill in the spaces where we make room for it is formidable.

The early autumn arrivals are noisy, and they come in massive waves, cackling and screeching like some demented hen party where everyone's had quite a bit too much to drink. The noisiest are pink-footed geese, birds that have now become a potent symbol for me of my time at Holkham. We had them at Raveningham, and they have been coming in increasing numbers to North Norfolk for just over a decade. Not far away, the new scrape on the area inland from the dunes known as Wroth's is almost ready. Earth movers and a phenomenally powerful rotary ditcher have ploughed up the ground so that it will collect water in varying depths. It will also be pocketed with little islets of vegetation. We used this machine, which employs a laser system to gauge the depth of the rotor – resembling a giant food-blender blade mounted horizontally like an outboard motor on a boat – and create what are known as 'foot drains'. These shallow scoops will hold standing water and create the best habitat for waders, including the lapwings, redshank and avocets that breed on the field. The rotary ditch digger is an expensive piece of equipment and costs about £25,000 for a hundred hours' work, but it has shifted about a thousand tonnes of mud. It's used by the RSPB, and Holkham obtained environmental grants (including a payment as part of our Higher Level Stewardship scheme) to fund it. The RSPB used the machine extensively to improve wading habitats on the new marshland reserve at Wallasea Island in Essex. Part of the island has been created using 3 million tonnes of waste earth excavated during the building of London's Crossrail project.

The new shallow channels at Wroth's will bring a fresh burst of life to the marshes. The geese will winter here, a respite from even harsher conditions in their native breeding grounds in Iceland and Greenland. The pink-feet come to feed in the arable farmland where there are stubble fields, sugar beet tops left over after harvest, which are a good source of carbohydrate, and the winter wheat. Yes, I truly love the spectacle they make, but I still don't have a favourite bird.

When I leave the village to get to my office, not far from Lady Anne's Drive, my route is along typically narrow, winding North Norfolk lanes. When I hit the coast road, I can see ahead towards the sea, which in turn gives its own special reflective qualities to that wide uninterrupted sky. The view is across the wetlands and the grazing fields where the belted Galloways are slowly, steadily, cropping the vegetation and churning up the soil, all the while making it more fertile for plants and insects. There's usually hardly another car on the road, and the sight of land and big sky makes me happy. I've not lived here that long, certainly not compared to the many years I spent further south at Raveningham, but now I've come to feel at home. The signs of approaching winter are here now and, out there behind the dunes, the natterjack toads are lying low, conserving their energy for the spring tumult when their thousands of voices will be raised for the spawning. Sitting at the Decoy, the dark line of the pine trees behind me shelters the marsh from the North Sea winds. On the other side of the creek, I can see movement on the ground as the vast flocks of pink-footed geese are readying for their morning flight. They begin to take off in waves, answering some rippling signal from their army of roost mates. As more and more birds rise against the pale sky, they make the air vibrate with their wings and send their wistful barking cry across the wide flat land. It's a scene of plenty, and a sign that all is not lost. Life will find its way back when we make space for it.

11

Cows and Calories

I'm often challenged when I say that we need to make space for nature. Within a minute or two (at most) of the conversation starting, someone will cry: 'But what about food security? If we don't grow our own food, we'll be vulnerable to outside threats!' Well, I've got news for you – in the UK we only grow 61 per cent of our current food, and we haven't been entirely self-sufficient since about 1850. However, we do still produce about 75 per cent of what Defra defines as 'indigenous food': think potatoes, cabbages, peas and turnips. Having said that, in the last thirty years, British vegetable production has declined from around 80 per cent of what we consume to just over 50 per cent.

It's worth understanding the concept of food security in greater detail. The Covid-19 pandemic, as I mentioned earlier, has made a lot of people wake up to the idea that the global supply chains on which we have all come to depend are vulnerable. Coupled with an increasing awareness that the government's plans for agriculture may result in fewer farmers, the crisis has made people nervous that a situation could arise in which our small islands might one day have to survive some sort of apocalypse. What if the only food available is what we can grow ourselves? All of these ideas and fears have become lumped in with a vague notion of 'food security'. It's not just about producing sufficient food that everyone in this country could survive without ever receiving any more food from any other source; governmental departments have defined food security as the nation's ability to ensure the availability of – and access to – 'affordable, safe and nutritious food for an active lifestyle, at all times'. Even today, the Food Standards Agency calculates that only 82 per cent of British households have access to all of the food they need. The rest, for economic reasons, have much lower levels of food security. Food security is one of the major social necessities believed to be under threat from climate change. On a global scale, food insecurity can arise from various factors: poor harvests, bans on exports for political reasons, rapidly rising energy prices, and so on.

Having access to sufficient food nationally is not the same thing as 'self-sufficiency', and producing food is not simply about making sure there is adequate land on which to grow it. It is also about ensuring that we have the technology and mechanisms that modern food production requires, for example things like fertilisers, pesticides, herbicides, animal feed and seeds. Many of these ingredients are not, and never have been, entirely produced within the UK. In the nineteenth century, as agriculture became more intensive, there was a big demand in this country for fertilisers containing ammonia and nitrates, much of it derived from collecting guano from islands in the tropics. Bird guano, for example, rich in nitrogen and phosphorous but also containing minerals like calcium, magnesium and zinc, became Peru's major export in the 1850s. An economic boom made the country rich from guano mining and exportation. Other sources were sought for this valuable commodity and Britain took some 300,000 tonnes of the odoriferous treasure from an island off the coast of south-west Africa in the space of just one year. There were conflicts even back then over the supplies, with 'Guano Wars' between Spain and Peru, Bolivia and Ecuador. Some guano is still extracted around the world today, and in the late 1950s Ian Fleming had his arch villain, Dr. No, prospecting for guano on the fictional island of Crab Key in the Caribbean.

We have long since replaced guano with other chemicals, of course, but if any of our supplies of the modern equivalents fail, or are cut off by political disruptions, then simply having the capacity to grow our own food may well not be enough. Currently we import almost 40 per cent of pesticide components, 55 per cent of our fertiliser ingredients and about a fifth of our animal feed. So, while the UK may be highly self-sufficient in food production, it may still depend on outside supplies and elements, without which things suddenly become difficult.

The Covid-19 pandemic should have taught us some practical lessons. While the UK seemingly had enough vaccines to immunise its own population, they could not be administered as quickly as had been hoped, because at one stage there was a worldwide shortage of the glass vials in which the vaccine was supplied. Aside from the very specific type of glass needed for medical applications, the Covid-19 crisis also caused a temporary shortage of the particular sand (silica) needed for the production of the vials. We live in a world where many things depend on global supply chains, over which we have little control when an unexpected event distorts the usual patterns. That is not to say, of course, that in a serious global supply crisis the UK could not increase its food production – as it did, dramatically, during the Second World War. It's not impossible that we might yet again have to 'Dig for Victory'!

Perhaps more serious than as-yet-unknown threats to global supply chains is the fact that, for many people, there is an acceptance of lifestyles in which excessive consumption has become the norm. This has obliterated any awareness of how our food is produced, where it comes from and how it is linked to the seasons of the year. The disconnection between the vast majority of people who live an urbanised life and those who work in the countryside and provide what we need from the land has become wider and wider. They see nothing strange in the notion of being able to eat lettuce in January or strawberries in November. 'Food security', in their minds, means eating avocados all year round, imported and flown in from places like Chile, Peru, South Africa and Spain. I'm not against avocados, but it's clear that shopping for food today is a very different experience from what it was for previous generations. I'm using avocados as an example because they were unheard of as supermarket produce until the 1960s, and for some years were a very marginal product, regarded as exotic and expensive in equal measure. We've got used to eating them though, and now import more than 100,000 tonnes of avocados each year, a quadrupling of consumption in the last decade. But would we starve without them? I think we know the answer.

Post-Brexit, the issue of how much food we import and export has become a hot topic. It's partly why the reforms to the way Britain treats its farmers have come into such sharp focus. When we were part of the EU, almost 70 per cent of the fruit and vegetables imported into the UK came from just two countries – the Netherlands and Spain. Meanwhile, 95 per cent of British lamb exports went to the EU along with 80 per cent of wheat exports. Aside from meat (and excluding lamb, which comes principally from New Zealand), the UK obtains only a small number of agricultural products from the EU, notably tea and coffee, soy beans, palm oil and sugar (both cane and beet). According to the NFU, in the years prior to Brexit more than 80 per cent of British beef went to the EU, and we also exported around 30 per cent of our lamb there. Naturally, government, consumers and food producers are watching what happens post-Brexit closely.

The topic of meat production has become a controversial one in recent years, attracting criticism because of the perception that it has a high carbon footprint and is more damaging to the environment than protein produced from plants. A lot of the most damning evidence for those claims comes from calculations based on the meat industry from American-style 'feed-lots' – a type of intensive production where the animals are essentially kept indoors and fed selectively with a diet which promotes rapid weight gain, rather than being allowed to graze naturally on pasture. Fortunately, this type of cattle farming is on a very small scale in the UK – currently. For this reason, and because of the peculiarities of our climate and latitude, the carbon footprint of British cattle farming is somewhat different.

In the run-up to Brexit, and indeed afterwards, the debate over how much of our food trade with the EU would continue, or from whence we might start to import other foodstuffs, became very heated. There were alarming stories about the prospect of chlorinated chicken or hormone-injected beef being imported from the USA, and other claims that meat would come from other countries where there were many fewer rules and regulations about how livestock were treated before they were slaughtered. Animal welfare, sustainability, regulations over what 'inputs' were allowed on crops and on animals in other non-EU countries fuelled the debate. In the middle of all this, British farmers argued that there would be no point in being rewarded for more sustainable food production if their crops and livestock became too expensive to compete with cheaper foreign imports.

In all the talk about producing food sustainably, and in

ways which safeguard nature and 'public goods', there is not much mention of how much people are prepared to pay for their groceries. According to a survey by the Rural Policy Group in 2021, the vast majority of people believe farmers should receive a fair price for what they produce, and almost as many of those surveyed said they would be prepared to pay more for their food to make that happen. The pressing environmental problems we all now face mean that paying a fairer price for our food may not be a matter of choice for very much longer.

Currently, by numerous measures British households are spending between 10 and 15 per cent of their household income on food. It's true that those on the lowest incomes spend a higher proportion of their income on food, but they still spend more on other things, including (unsurprisingly) housing, but also on transport and even recreation and leisure, according to official figures.

This access to cheap nutrition is a relatively recent phenomenon. In the eighteenth century, working people spent about three-quarters of their income on food – and the highest value item was always bread. Even as recently as the 1950s we were spending a third of our income on food. No one wants to go back to the living conditions that prevailed before the Victorians, but the fact is that food has never been cheaper in this country. A broiler chicken, which would feed four people, is now available for as little as £3.50, and sometimes less, but certainly much cheaper than a pint of beer or barely more than a coffee in London. The UK eats its way through over a billion broiler chickens per year. I can remember when my mother made a roast chicken for a Saturday lunch as a treat, and the carcase would later be used to make soup for our family. In real terms, chickens were at least three times as expensive when I was a child. For this reason, we continue to generate vast amounts of 'food waste'. According to the charity WRAP (Waste and Resources Action Plan), more than 35 per cent of greenhouse gases in the UK come from food production and consumption. They claim that a more efficient use of our food resources would cut CO_2 emissions by 100 million tonnes over a decade.

There is some good news: in the last few years, much of the waste food production in the UK has been reduced, but it still adds up to around ten million tonnes each year. The Intergovernmental Panel on Climate Change estimates that, worldwide, at least a quarter of the food produced is wasted, possibly more, and that this waste contributes 10 per cent of the world's man-made greenhouse gases. More than a quarter of the food grown in the UK is never eaten, a quantity that accounts for more than 5 per cent of the nation's greenhouse gas emissions. That also represents a colossal waste of energy, land and water.

The Government Chief Scientific Adviser, Sir Patrick Vallance, has stated publicly that climate change is a bigger threat to humanity than Covid-19, and says that changes in how we all behave – and, especially, in what we eat – will be inevitable if we are to stave off the worst effects of that change. The pandemic also prompted many people to say they recognise that changes in behaviour on all sorts of fronts need to happen, with renewed interest in sustainability and food security, and a perceived wish to buy 'local'. Meanwhile, as part of its net zero strategy, the government has said that carbon taxes – in other words, levies on food with a 'high emission footprint' may be necessary, even though they also state that there are no plans for such taxes . . . as yet. It's not surprising that people are confused. However, the evidence for why we need to think carefully about which types of food we produce and consume is not so confusing.

The 2021 Dimbleby report on the UK's food strategy states simply that by 2032, the national consumption of fruit and vegetables will need to rise by 32 per cent and of fibre by 50 per cent, and that meat consumption will need to decline by 30 per cent, with a similar decline in the amount of HFSS foods (High in Fat, Salts and Sugars) we consume. This is not just about the environmental impact of what we produce. There are other hugely costly reasons the government wants the way we eat to change too – namely the statistics that show that almost a third of adults in the UK are obese, and another third 'overweight'. Perhaps more worryingly, a fifth of children are either obese or overweight by the time they start primary school – aged five. The healthcare burden that obesity is likely to place on the state is enormous. The government estimates that across the UK the NHS costs attributable to obesity will be about £10 billion per year by 2050, with wider societal costs (including reductions in economic development) estimated to reach £50 billion.

The boffins who have analysed our agricultural system have made some startling calculations. One of them shows that if we turned just 9 per cent of our least productive farmland over to nature then we would only be producing 1 per cent fewer calories out of our total national consumption. Turning 21 per cent of that low productivity land over to nature would reduce our calorie production by just 3 per cent. If the boffins have got it right, that still leaves the UK producing 99 per cent of its current crops, 97 per cent of current fruit and vegetables and 99 per cent of our milk, chicken and pork. The only sectors which see a relatively significant decline under this scenario are beef production, which would go down to 92 per cent of current levels, and lamb, which would reduce to around 80 per cent. These figures assume greater significance when you factor in the calculation that more than 80 per cent of the farmland in the UK is devoted to livestock – and yet meat, dairy products and eggs account for less than a third of our calorific intake. Almost 70 per cent of our calories come from plants, and that uses only 15 per cent of our farmland.

A shift in our diets towards more plant-based content would use our farmland more efficiently. That would make space for nature, sequester more carbon and align with the government's professed wish to promote the good of our long-term health. That doesn't seem too radical, especially if it allows us to restore nature to those parts of the country that are taken out of intensive production. The government has said that its *Path to Sustainable Farming* plans acknowledge what they call their '30by30' target: a pledge to protect 30 per cent of England's land for biodiversity by 2030 through habitat creation and restoration. By some measures, more than 25 per cent of that land already enjoys some level of protection, although some of it is rather toothless. We must also remember that almost all of that 'protected' land is still actively farmed. So, the alarmist claims that the 30by30 plans will endanger food production need closer examination. Wildlife organisations claim that less than 10 per cent of the country enjoys statutory protection, and much of that is poorly policed. However, it doesn't seem unreasonable, when we think about how farming can contribute to this better, healthier, less damaging view of the world, that we start valuing our agriculture system more intelligently.

The fact that we have plans which seek to improve our health, our climate and our natural environment, must surely be a good thing. The British countryside, and how we farm it, remains at the heart of it all. Instead of removing farmers from these discussions, it seems to me that these sorts of calculations put them at the heart of it. They, more than any other group, have the power and the ability to reduce carbon emissions, sequester carbon and improve the natural systems that sustain it all. Placing greater value on the food they grow and valuing the types of food they can produce for our benefit, reinforces their importance to the overall functioning of our society. The plan envisaged by the scientists advising the government on the future of farming is that, while some farms will effectively go out of business, and some land restored to native woodland or species-rich meadows, there will also be space for farms which produce higher yields and take advantage of improvements in technology, including chemicals and selective breeding of species (both plant and animal), to maintain our food security without destroying all that nature provides. What farmers are not getting right now, it seems to me, is a coherent plan that allows them to work out how best to deliver on these things. If farming and farmers don't have a clear plan for what's next, and believe they can't afford to stay within the industry, then all the talk of a better future, a more sustainable type of agriculture and a healthier planet goes out of the window. But is this the fault of the farmers?

There's been a lot of talk about 'rewilding', and about 'sustainable' or 'regenerative agriculture'. People use the terms interchangeably, although there are significant differences in the approaches. According to the National Food Strategy, any future 'food system' should 'make us well instead of sick', be resilient to global shocks, help restore nature and halt climate change 'so that we hand on a healthier planet to our children', and meet the 'standards the public expect' on health, environment and animal welfare. Achieving this, according to the strategy, means more of the countryside being used to sequester carbon and restore nature. This will require 'diverse methods of land management'.

Making space for nature doesn't have to mean that farming becomes less cost effective. Lower-carbon, more sustainable farming which provides us with a healthy diet need not be more expensive. But the scale of these tasks is huge. If the UK is to achieve its legally mandated 'net zero by 2050' target (set out in the Climate Change Act of 2008 and subsequently amended in 2019), then the emissions from farming alone will have to be cut in half. To achieve this, forests must be planted and grown to sequester any remaining emissions, and habitats like peat bogs and marshes must be restored to create more carbon sinks. Again, these are ideas that make some people nervous. Those people might be less anxious if they understood that currently only 15 per cent of the UK's total land 'footprint' is used for growing key staples like potatoes, fruits and vegetables. The 'land footprint' is not just land in the UK, but also includes the land overseas used to produce the food we consume. If we use beef and lamb production as an example, that accounts for 40 per cent of the UK's land area in addition to a land area overseas which is three times larger than Wales.

Models have been produced which look at how the UK's farming sector could change to address the urgent issues surrounding reducing carbon emissions and improving the state of the natural environment. The government has considered three scenarios: 'Intensifying' agricultural production while turning over some land to environmental schemes; increasing 'Agroecology' (farming using the principles of ecology); or introducing 'Resource Efficiency'. They looked at the effects these three models would have on food production, greenhouse gas emissions and biodiversity.

Under the 'Resource Efficiency' model, synthetic inputs – chemicals – were reduced, and livestock management was integrated more sympathetically into the crop rotations. This approach resulted in a need to shift production into the growing of more vegetables and pulses, resulting in the same agricultural 'output' but greater biodiversity overall on the farm. The 'Agroecology' model involved reducing livestock density, eliminating the use of chemicals and restricting the use of antibiotics for livestock. This resulted in lower levels of food production, almost the same level of greenhouse gas emissions but potentially higher emissions, if farm output resulted in higher levels of imports. However, it led to higher biodiversity on the farm. Interestingly, the 'Intensification' model led to keeping the production of the farm stable, even though a proportion of the land was turned over to nature. It resulted in lower biodiversity on the farm itself, but greater biodiversity overall on the land given back to nature. It also produced a reduction in greenhouse gas emissions. It seems to me that some kind of combination of these strategies is likely to be a realistic way forward. As with most things to do with how we use the land for farming and for nature, there needs to be some tailoring. One size does not fit all.

The Burrows
November

Driving out of the hamlet of Warham, I cross the little bridge over the Stiffkey. Near the bridge, an arable field slopes downhill, allowing whatever nutrients are put onto the land to drain into the river. In the Burrows field, the wildflowers are all gone now and there is a sea of dead brown seed heads where the grasshoppers, snails, brown skippers and Burnet moths clung so plentifully in high summer. It's raining steadily, and there is a stiff easterly wind that drives the pellets of water in stinging bursts against my face. The sheep are back in the field now, grazing steadily, heads down and impervious to the weather. In the adjoining field, where the fort stands high above the river valley, a small group of cattle are taking shelter under a veteran ash tree, tucked against the hedge boundary.

The tree is well over a hundred years old, and the whole scene is picturesque. But a closer look reveals that it has shed some significant branches, and the leaves are showing signs of dieback, the chronic fungal disease caused by *Hymenoscyphus fraxineus* (sometimes called 'chalara'), which has swept through the majority of our national population. The disease was first spotted in England in 2010, having moved westwards from Poland after being somehow imported from Japan. Up to 70 per cent of affected trees will die, with saplings and younger trees succumbing quickest. The older trees, like this one, can withstand the fungus, but it weakens them nonetheless and if another pest, or another disease like honey fungus strikes, then the cumulative stress is often fatal. Perhaps this one is a survivor. The cows stare at me, curious but calm, and show no wish to leave the shelter of the old ash. They have turned the track into the fort into a slippery mud mush, and there are copious cowpats all over the narrow entrance that leads into the central doughnut. I recall the thousands of chalkhill blue butterflies and the grass carpeted with spider webs of just a few months ago. Up on the outer wall, the wind is gusting strongly and I turn my face away from the rain as best I can.

The fort and the surrounding fields cover about a hundred hectares here, once again similar in size to Great Farm. On the near horizon I can see the outline of St Mary Magdalene with its Saxon tower. The church has been tinkered with, rebuilt and remodelled by the Normans and those who came later too. Some elements have been removed, and others added, over the 900 years of its existence. There may well have been a wooden church on the site that was even older. Change within the landscape is just as possible, and as climate change progresses it seems entirely likely that we will start to notice fluctuations in the way the land responds to the shifts in the seasons. If sea levels rise and coastal erosion increases, should we be planning ahead for that by creating habitats for the birds and animals which may be displaced further inland? It seems to me that the land can also be remodelled, in tune with the needs of the time.

Coke of Norfolk was not afraid of a bit of landscape remodelling, if it would benefit the farm. He was, as the historian Susanna Wade Martins points out in her numerous papers on Holkham, an irrigation enthusiast. In 1803, through the Norfolk Agricultural Society, he began to put up an annual prize consisting of 'a piece of plate to the value of 5 guineas to such a person as shall convert the greatest area of waste or unimproved meadows into water meadows in the most complete manner'. Coke encouraged the creation of a new system of water meadows along the River Nar at the beginning of the nineteenth century, land he owned at Castle Acre.

We can drain land, or we can flood it. We can put crops on it, or let it revert to scrub. If we wanted to, we could even remove some of the hedges here, many of them planted at the time of the Inclosure Acts. Here, around the Stiffkey, we could use the dykes and drains to allow the river to flood periodically, meandering gently through that expanding alder carr habitat. Looking down onto the floodplain I would be able to see spoonbills and lapwing breeding, bringing life and change to what has been a fairly impoverished farming landscape for many years. This fort would be a viewpoint, something we lack in this part of Norfolk, even though the land hereabouts is not as flat as people think it is. Holkham is actually part of what they call High Norfolk, and the sandy loam over chalk rises in gentle undulations. We could bring back many birds and butterflies and insects right here. I'm not sure we should be bringing in the beavers just yet. I'd rather let the landscape do what it wants to do, adapting to the positive changes we encourage. All without losing its potential to produce food.

This is the barren season, and there is little colour in the land, but the carr stands out, a strip of dark green to the west across the river. Nearer to the fort and skirting its northern edge, the river is invisible, cloaked in vegetation. The sheep are grazing the wildflower meadow to my east, and southwards I see the low-lying fields between the fort and the small village of Wighton. A screen of straggly poplars rises high from the edge of the village. As I take all of this in, and in spite of the grim conditions, I am fired up by the idea of how simply this area could be transformed. By digging out a few side channels north-west and south-west of the river, it should be possible to turn the fields nearby back into the marshy floodplain that it once was. It would be relatively inexpensive, and it would return the river valley to something much closer to its original flow. It would revert to how it was before it was diverted with the man-made embankment in an attempt to drain the valley for crops. The habitats that this created would be more dynamic, and it would create 'edge'. It would very quickly make these fields attractive to waders during the breeding season and increase biodiversity significantly. I imagine hundreds of lapwings in front of where the poplars are, and waterfowl using the banks of the chalk stream in the spring.

When I look out over the bend in the river, I see all kinds of possibilities. It's the kind of landscape-scale change that is even more exciting than trying to convince landowners to sequester portions of individual fields to increase biodiversity. This land, returned to its floodplain habitat, would be an asset to the community, and a prime example of natural capital restoration and protection. As farmland, in spite of the drainage, it is of very little productive value, and as an asset to the landowner it only increases in value if he sells it. The fields across the river are no good for grazing as they are full of rushes and reeds – plants that love water. The land is low-lying and boggy. It gets topped once a year at most, perhaps less often. The woodland is criss-crossed with drainage channels, but you can see that the land wants to revert to its natural, boggy state.

Right now, it's poor value on every front, probably only earning anything from the Basic Payment Scheme. But when I look at the fort and the river and all that surrounds it, I just see possibilities. This is land that I know I can heal. It should be as high in natural capital asset value as the fort and the wildflower field beside it, where I've put the sheep to graze to the north. If the river can be properly restored, the floodplains opened up and the alder carr reinstated, then this valley should have even more biodiversity value than the protected sites. In terms of natural capital, it would provide clean river water, waterfowl nesting sites and historic interest, and thereby become a greater public asset. People in Wighton would look out over a diverse, healthy habitat and yet, with some permitted light grazing, it could still be part of a farmed landscape. It wouldn't be turned over to nature and simply left, because it would still be providing food for animals, but it would be managed so that the sheep or cattle did just enough to make the sward and the meadows healthier for other species. We would be resetting the clock. To borrow a phrase from the rewilding enthusiasts, cattle can be ecosystem drivers.

There is money available under existing schemes to do some of this work, and I know that the government's Green Recovery Challenge Fund has earmarked money for the chalk stream project. The fund was set up during the pandemic to help the UK 'build back greener', pledging a total of £80 million which would be allocated in grants of anything from £50,000 to £5 million for almost a hundred projects that promise to restore nature and help address climate change. It's seen as part of a broader green recovery which aims to increase jobs and skills in the environmental sector, especially where they provide what are called 'nature-based solutions to climate change'. Projects are favoured which prioritise conservation and restoration, that demonstrate nature-based solutions to climate change mitigation through planting trees, and also connect people with nature. In the case of the Stiffkey river valley, it's a 'tick, tick, tick'. As well as stimulating natural recovery in the landscape, the scheme is also about providing up to three thousand jobs, and training young people in skills that include ecology, surveying and education, as well as supporting businesses that supply things such as agricultural engineering, equipment and seeds.

The money for this 'green recovery' is being awarded subject to approval by Defra through a partnership between the Environment Agency, the National Lottery and Natural England. Under this scheme, the Norfolk Rivers Trust and the Norfolk Coast Partnership were successful in applying for funding to restore the catchment areas of the Stiffkey and the Hun rivers. Like the Stiffkey, the Hun is a chalk stream – although it is much shorter, running for just under four miles near Hunstanton, west of Holkham, and emptying into the North Sea at the Holme Dunes National Nature Reserve. This is a rich habitat and, like Holkham, is home to natterjack toads, otters and a profusion of invertebrates. By contrast, the Stiffkey is eighteen miles long, and the Norfolk Rivers Trust has already been successful in applying for additional funding to explore setting up an 'environmental impact bond' which would aim to reduce pollutants, especially phosphates, from entering the river. I strongly believe that these kinds of partnerships, aided by the private sector, may be key to regenerating a healthier countryside. There are ways to involve the private sector in these landscape-scale restoration and improvement projects which don't involve profiteering from the countryside. More and more big companies want to be seen to be doing their bit for CSR (Corporate Social Responsibility), and the river valley around Warham's fort would be an ideal project of that sort. I think it's entirely feasible that a company, or a pension fund, would be willing to sign up to a thirty-year project to restore this patchwork of fields, the river and the ancient alder carr growth.

12

Grass Is Good

The biggest single crop in England is not wheat, or barley or potatoes: it's grass. Pasture and semi-natural grassland covers almost 40 per cent of our agricultural land. That's about four times as much as the total urban and developed area, and more than twice the amount of land covered by permanent crops. Pasturage is highest in the south-west and the north, and lowest in the east (where arable crops dominate). And grass, like other plants in the agricultural landscape, is made up of diverse species, although the range of grasses used for pasture now is smaller than it once was. Defra advises that grazing should begin when the grass has reached between 8 and 12 centimetres, and that the average sward height should be no less than 5 centimetres – with at least a fifth of it over 10 centimetres to provide invertebrate habitat, and 10 per cent of it less than 10 centimetres high so that birds can find food. A field of grass, you see, is not just a field of grass; and the experts will be out measuring it with their sward sticks while also assessing its moisture content when it's time to cut the hay. Grass provides total food for about three-quarters of the UK's ten million cattle, and almost all of the feed for our thirty-five million sheep.

Recently, I took one of the country's largest landowners for a tour of Holkham Estate. This prominent individual was interested in nature, and fascinated to see how we managed to keep livestock on the marshes alongside such healthy bird populations. 'How will I know if I'm getting things right?' he asked. For me, the answer is relatively obvious: it's the colour of the grass that tells me if a field is under too much or too little pressure. It may be temporary, permanent or rough grazing. In winter it may turn golden and wispy, in spring the fresh new shoots will appear, and in summer it will be vibrant and green. Ideally, we aim for 'herb-rich leys' – in other words, grass with some other species in it – although when grazed by cattle the variety of plants diminishes to an extent, less so with sheep. Grass is an amazing plant, and it will put on growth as soon as the soil reaches about 6° C, which can increasingly happen in the winter. At Micklefleet, close to Lady Anne's Drive, I saw massive flocks of widgeon – more than two thousand birds – taking advantage of fresh growth after a couple of weeks of mild winter weather. After the birds had been on the field for about ten days the fresh growth had been removed – but the spring will see the return of the cattle, which we will move around the estate making the most of the more vigorous growth that comes with warmer days.

I think that many people don't understand that without the grazing animals, primarily cattle and sheep, the species diversity in a lot of our countryside would be lower. At Raveningham we had a field known as the Hundred Acres. Ironically, it was a very small area – not much more than a single acre – and it was left alone until late summer every year when it was cropped for silage. Next door, there was a larger area of pasture and occasionally a few dairy cows would move through the Hundred Acres to graze it. We carried out insect surveys and found that it contained a very high number of moths and butterflies. And then the decision was made to remove the cattle from that part of the estate as they had become uneconomical. Two years after they had gone, we surveyed the Hundred Acres and found that it was greatly impoverished. Many of the plants and invertebrates that had always been seen there had disappeared.

In all the talk about how to make our farming practices sustainable, livestock comes in for a lot of stick. We have just over nine million cattle in the UK, around 75 per cent of them cows. Dairy cows (55 per cent) outnumber cattle bred for beef (45 per cent). Meat, and especially beef, production is often cited as one of the most damaging aspects of agricultural production, especially when it comes to greenhouse gas emissions and the carbon footprint of cattle. In food terms, they say, beef emits thirty-two times as much CO_2 as tofu. According to the UN's Intergovernmental Panel on Climate Change, the global food system is responsible for anything from 20 to 37 per cent of all greenhouse gas emissions. They classify this as AFOLU (Agriculture, Forestry and Other

Land Use), so the numbers need to be unpicked a bit. They acknowledge that AFOLU results in both emission and removal of greenhouse gases. A proportion of emissions are removed – or offset – from the atmosphere by the carbon stored in soil, in living biomass and dead organic matter, although just how much is difficult to calculate. What's more certain is that animal-based foods are responsible for more than half of greenhouse gas emissions from agriculture, and include CO_2, methane (CH_4) and nitrous oxide (N_2O). Road vehicles, aeroplanes, trains and ships produce about 16 per cent of the total emissions. The other big emitter is, of course, energy production, very predominantly from oil and gas, accounting for about 25 per cent of the total, while industry comes close behind at 21 per cent.

Each of the so-called greenhouse gases is allotted a GWP factor (Global Warming Potential) based on how much energy 1 tonne of a gas will absorb over time relative to the emissions of 1 tonne of carbon dioxide, over a period of 100 years. CO_2 is the base reference with a GWP of 1, while methane is given a rate of 25–36, although it stays in the atmosphere for about a decade rather than a century. Nitrous oxide is worse, with a GWP of around 300, and the fluorocarbons are many, many times higher.

Currently, the world produces the equivalent of around 40 gigatonnes of CO_2 emissions per annum. A gigatonne is 1 billion tonnes, and helpfully equates to the mass of all the land animals on Earth – excluding people (who altogether weigh about half a gigatonne).

Cattle produce more than 4 gigatonnes of CO_2, while cow's milk on its own emits another 1.6 gigatonnes. Beef and cow's milk are especially carbon intensive, responsible for more than a quarter of all the agricultural emissions. Cows notoriously also emit methane at both ends, mostly from 'burping' as their digestion converts plant sugars into simpler molecules for absorption into the bloodstream. Atmospheric methane (which is two-and-a-half-times higher now than before the Industrial Revolution) is emitted from the oil and gas industry too, but since it is, by some measures, at least twenty-five times more potent as a warming element in our atmosphere than CO_2, then anything which reduces it must be a useful weapon in fighting emissions. And as a proportion of livestock emissions, methane is around 40 per cent.

At COP26, I listened as more than ninety countries committed to reduce methane emissions by 30 per cent by 2030. We need to remember that, aside from cow burps, methane comes from oil and gas wells, leaks in pipelines and other equipment, and from decaying matter in landfill. The International Energy Association calculates that global methane emissions are in the order of 570 million tonnes, about 60 per cent of them from human causes. Energy production and agriculture are the two biggest culprits. The two largest producers of methane from gas and oil are Russia and the USA, accounting for about half of the industry totals. The IEA says that to reach net zero by 2050 methane emissions from fossil fuel operations will need to be cut by 75 per cent.

It's all very well spouting these big numbers. To put them into perspective, we need to understand that one cow can emit up to 500 litres of methane per day, but we're learning that there are some fairly simple ways in which that can be reduced, including by adding things like garlic, linseed oil and seaweed to cattle feed, which are all showing promising results. But, regardless of these helpful improvements, as the world population grows, its appetite for meat is also incresasing. The way livestock, especially cattle, are kept will affect how much methane they produce. However, some stark facts need to be faced. Producing 100 grams of protein from beef emits an average of 25 kilograms of CO_2. The lowest figure is less than 10 kilograms per 100 grams of beef, but up to 50 kilograms of CO_2 per 100 grams for the most wasteful types of husbandry. The equivalent figures for pork are 6.5 kilograms of CO_2, and just over 4 kilograms for 100 grams of chicken. The figures for protein derived from growing green vegetables or nuts are significantly lower, usually much less than 1 kilogram of CO_2 emissions per 100 grams of protein.

The other problem with cattle is that when they are kept under intensive conditions they may be fed with ingredients such as soy protein, which might itself have been produced in an environmentally damaging way, such as by converting tropical rainforest into cropping areas, notoriously in the Amazon basin. In other places, even where the cattle are put out to graze their pasture may have been created using intensive chemical fertilisation. In Britain, beef cattle eat mostly grass, meaning they are not drivers of deforestation in the tropics. Some supplementary feeding with cereals enables them to produce protein more efficiently. And unlike cattle kept in some beef-producing countries, the British herds consume water naturally available from rainfall.

Sometimes, though, a closer look at the numbers yields results which make for uncomfortable reading. For example, while cattle farming in Britain produces lower emissions than in many other countries, such as Ireland or Australia, it does not do as well as France, Germany or Denmark. CO_2 emissions by weight of protein produced are also lower in the USA, but this is because the intensively reared cows are fed on grain and treated with antibiotics and growth hormones so that they grow faster and can be slaughtered at a younger age. The harsh reality is that those cattle therefore spend less time alive and less time emitting methane. However, we may feel moral qualms about subjecting cattle to a life lived almost entirely indoors. I like to see cattle out in the field, rather than penned inside.

One of my mottoes is 'Think local, act global'. If you're feeding your cattle soy grown in South America, perhaps you need to think again. Yes, cattle fed 'naturally' on grass take longer to fatten up until they are ready for slaughter, but putting cattle onto the land – as we have shown at Holkham – has other, positive, knock-on benefits. The way they graze and affect the soil structure through the variety and density of vegetation, and so on, may well be having positive environmental benefits. And land which is kept as permanent pasture is covered in grasses which grow, regenerate and hold carbon. While a tree will store carbon in its leaves and wood while it is growing, it releases that carbon back into the atmosphere when it is cut down and burned, for example. Trees also take much longer to reach their best carbon-storage potential, although they perform many other ecosystem services while they are growing. However, grass quickly stores carbon in its roots and then underground in the soil. That carbon is not affected by what happens to the leaves above ground. Depending on size and the number of animals per acre, farms which feed their cattle on grass pasture may in fact be carbon neutral.

A typical carbon audit of a mixed farm will often produce data that says the business is 'carbon negative', usually because of cattle, but it doesn't take account of the biodiversity that grazing encourages or the carbon stored in grassland. Pasture isn't normally ploughed – another way in which carbon is released into the atmosphere. Grazed pasture will also require less fertiliser and fewer pesticides or herbicides. According to the NFU, British beef and lamb provide some of the most efficient and sustainably produced meat in the world. They say that, of the total of around 50 million tonnes of CO_2 emitted by the UK, agriculture accounts for around 10 per cent – about 45 million tonnes. Cattle and sheep account for more than half of that again, almost 6 per cent of the UK's total emissions. However, the NFU says that more than 9 million tonnes of carbon is sequestered in grasslands, and that this reduces the total of carbon emissions, making cattle and sheep account for less than 4 per cent of the UK total.

To reap these benefits, you need to keep fewer cows on your land, and this inevitably may mean the farmer makes less money from his herd. Once again, the solution, or at least part of it, is to educate the consumer. We must show them that instead of buying the very cheapest type of meat available in the supermarket, it is worth eating better quality meat – and less often – to keep farmers in business and not kill the planet. As I've said before, a lot of farming production is driven by the buyers higher up the food supply chain. The popularity of the Aberdeen Angus is largely due to its ability to reach the right dead weight in a shorter period of time than other breeds; it probably produces a carcase with a dead weight of 250kg four or five months earlier than Holkham's Belted Galloways. The Angus beef is perfectly good, but it was originally promoted because it was more economical to produce, rather than for its superlative taste.

Cattle aren't just about meat, of course. The NFU says that British milk production is much more efficient than in many other countries, producing only about 40 per cent of the global average emissions for the industry. If that level of efficiency were replicated worldwide then we could produce the same quantities of milk with about a quarter of the dairy herds currently kept around the world. An IPCC (Intergovernmental Panel on Climate Change) report in 2019 calculated that one litre of British milk had a carbon footprint of

1.2 kilograms compared to a global average of 2.9 kilograms. What I'm saying is that we need to be careful when we listen to blanket condemnations about the damage beef and dairy farming in this country do to the environment.

If we measure all of the land used to produce food for the UK, and that includes the land area used overseas to produce our food, then 85 per cent of it is used for rearing animals. This huge swathe of land produces just 32 per cent of our total calories. Using the same basis for calculations (as described in the 2021 Dimbleby food strategy review), we use 15 per cent of our land footprint to produce 68 per cent of our calories, and this is all from plants. And the UK's Climate Change Committee is recommending that, to reach net zero by 2050, we need to reduce the amount of meat in our diet by 20–50 per cent.

These figures don't account for what the government calls the 'opportunity cost' of livestock farming. This refers to the potential that the land used for livestock would have to capture carbon if it were not being used in that way. They believe that we would go a long way to meeting the net zero targets if we removed the least productive 20 per cent of land from agriculture and planted it with broadleaf woodland, or restored peat bogs.

It is these sorts of numbers that inform the government's scientific view that a switch to greater quantities of plant-based foods is a key weapon in the fight to reduce the national carbon footprint and stave off the continuing rise in global temperatures. Repeated studies have demonstrated that the most processed and energy rich food is worse for the environment by several metrics. It tends also to be less nutritious and more likely to fall into the HFSS (High in Fat, Sugar and Salt) categories.

One of the factors that often goes unacknowledged in the low-carbon plans aimed at making us eat more healthily is that, at the moment, animal protein is a cheaper alternative to plant-based food for many people. Aside from this, in some cases, commercially produced non-meat products may contain more processed ingredients than farmed beef or lamb. As market share grows, it's hoped that plant-based staples will become more competitive. Supermarkets are already making many 'own brand' versions of the first meat-substitute products that appeared on the market. However, as things stand, for example, cheap beef mince is less than half the price of soy substitute 'mince'. If a consumer wants to replace a meat-based ready meal with a 'non-meat' ready meal, they may well balk at the price.

Currently, there is an emerging market for what are called 'meat-mimicking products'. There are 'burgers' made of soy or other plant materials which are shaped, coloured and textured to resemble a beef burger. They tend, on the whole, to be more expensive than the products made from actual meat. The National Food Strategy suggests that growing the crops needed to expand the production of plant-based products will create significant numbers of new jobs, both in manufacturing and in farming.

The bulk of the UK's current food consumption is

still a mixture of processed foods and red meat with a high protein and calorific value. We have the highest rate of consumption of ultra-processed foods in Europe at just over 50 per cent (followed closely by Germany and Ireland). Producing these foods has, as I hope I've shown by now, come at a high cost for our natural environment, and with a high output of greenhouse gases. Often, they have also been produced relatively inefficiently, in terms of use of available land – and according to government statistics, eating this diet has come with significant health costs. Around 75 per cent of British adults between forty-five and seventy-four years of age are either overweight or obese. According to the National Child Measurement Program, by the age of eleven – the start of secondary school – about 30 per cent of our children are overweight or obese. This translates into just over one million annual hospital admissions for conditions associated with obesity.

The science tells us that we don't depend on animal protein to survive: we can obtain the same amounts of protein from plants, and significantly reduce greenhouse gas emissions. At the same time, farmed in the right way, these ingredients would allow us to increase the biodiversity on our farmland while improving our overall health and almost certainly reducing the rates of national obesity. There is currently, however, a drawback: the cost to the consumer, and a lack of awareness about how to cook and prepare a more heavily plant-based diet.

Vegetarian 'sausages' are at least twice as expensive as pork, and oat milk is about 20 per cent more than cow's milk. Even if we ignore the cultural barriers to making a switch in diets, and although the price differentials are becoming smaller, for people on a limited budget, the perception is still very much that buying processed vegetarian or vegan alternatives is too expensive. In the current market, consumers are also paying a premium for products that are convenience foods, saving time and effort when preparing a meal, even if the same meal could be made much more economically by purchasing the raw ingredients.

A survey of some meat-mimicking products in major UK supermarkets found that the price differential was anything between 10 per cent and more than 200 per cent higher than those containing real meat. A meat-free spaghetti bolognese, for example, was almost double the price of a 'real' spaghetti bolognese ready meal. However, the healthier alternative, a bolognese using lentils and made at home from scratch, cost 15 per cent less than the meat version. Other typical meals showed even starker differences – a chicken curry was half the price of a commercial tofu-based product, but a home-made vegetarian curry with high protein content would cost just a third of the meat-based version. As ever, when talking about costs it is worth remembering that traditional meat products may well be produced without factoring in the hidden costs of the carbon footprint of production. No government yet seems ready to pass on those 'carbon costs of food production' to the consumer. That may have to change.

As I've said many times, I'm not a campaigner for

vegetarianism – and if someone told me I had to give up meat, they'd get short shrift. But, reducing the amount of meat we consume and learning to adopt some healthier dietary habits are significant factors in how we make our current farming systems better for the environment. I do believe that we are at a turning point. I've tried synthetic meat (which is in itself highly processed) and found it palatable, although for me the texture is not comparable to real beef. Recently, I attended a dinner where we were served jellyfish for starters as a sustainable alternative to fish, as well as chocolate made without cacao and coffee made without coffee beans – which didn't really work for me.

There are lots of positive things happening, and more and more farmers are actively trying to do the right kinds of things on their land. But I'm also nervous. The Environmental Land Management schemes, which were designed to replace the Common Agricultural Policy subsidies that provided the backbone of many farmer's incomes during our time within the EU, seem now to be unambitious. I'm really not convinced they will deliver the aims of the 25-year Environment Plan. The plan stated that the government wanted to ensure that resources from nature – food, fish and timber – would be used more efficiently and sustainably. What they termed 'resource productivity' would be doubled by 2050. By 2030, they said, we would be improving the management of our soil with 'natural capital thinking' applied to develop soil metrics and management principles. Food, it claimed, would be produced both sustainably and profitably.

As things stand, the proposed ELM payment rates are not high enough for farmers and land managers to apply the suggested schemes at significant enough levels. The Sustainable Farming Incentive (SFI) is scheduled to open in 2022 to all farmers registered with the Rural Payments Agency. It was only in December 2021 that Defra clarified that the payments made under the new schemes – or standards – would have three 'levels of ambition', with the introductory level setting out to pay farmers to meet what it called 'a good level of sustainable environmental practice alongside food production beyond the regulatory baseline'. Higher levels of ambition will receive higher payments and 'will include more challenging actions and achieve higher levels of impact on the environment and climate change'. The payments will last for three years at a time, and farmers can choose to enter the schemes at what the government calls 'field level', i.e. they can choose specific fields to enter into the SFI programmes. There is no minimum or maximum area of land that can be entered into the SFI. Just like the Basic Payment Scheme, the farmer must hold a minimum of 5 hectares of eligible land – but not all of it needs to be entered into the scheme.

One of the hopeful signs in the new system is that there is also a commitment to be more flexible on the part of Defra when it comes to encouraging farmers to remain within these schemes, and the government says it will assume 'good faith' on the part of participants. One key difference from the former EU system is that farmers who fail to meet the standards will receive advice on how to improve. Failure to meet the objectives may see payments reduced, but farmers will not be fined or have financial penalties imposed.

The details of SFI revealed in December 2021 reiterated that farmers will be paid for 'public goods . . . such as water quality, biodiversity, animal health and welfare and climate change mitigation, alongside food production'. In 2023, farmers will also be able to apply for funding under the Local Nature Recovery and Landscape Recovery schemes. Landscape Recovery will undergo a pilot phase this year (2022) designed to support long term, large scale land-use change and habitat restoration projects. Defra's hope is that 70 per cent of English farmers will sign up for these agri-environment schemes by 2028, and that's more than double the current level of participation. I applaud the ambition of plans which are aimed at encouraging farming practices to evolve, rather than simply penalising those who don't want to participate.

In the grand scheme of things, the government has high ambitions. The SFI scheme's soil standards could save some 60,000 tonnes of CO_2 (equivalent) in 2023–2027. In 2033–2037, the schemes could deliver carbon savings of 800,000 tonnes in England, the equivalent of taking 400,000 cars off the road during *each* of those years. There are three standards envisaged – for Arable and Horticultural Soils, Improved Grassland, and Moorland and Rough Grazing. Most farmers in England will be eligible to take advantage of the soils and moorland standards, which will bring considerable benefits for both carbon emission reductions and increased biodiversity.

There is much technical detail to the standards, and farmers must ensure that by implementing them they don't cause adverse effects upon sites of special scientific interest or affect scheduled monuments. At the Introductory Level, payments for arable soils will start at £22 per hectare and involve preparing a soil management plan, testing for organic matter and providing 70 per cent cover over winter. At Intermediate Level the payment increases to £40 per hectare and the cover over at least 20 per cent of the land must be multi-species. The Advanced Level standards have not yet been finalised, but Defra says that a degree of 'no tillage' is likely to be incorporated. Similar plans are laid out for the Grassland improvements, including the establishment of 'diverse sward with a mixture of grasses, legumes, herbs and wildflowers on at least 15 per cent of land entered into the standard'. The payment rates are slightly higher than for arable soils, i.e. £28 per hectare at Introductory Level, and £58 per hectare for Intermediate. One key provision is that no more than 5 per cent of soil may be left bare over winter.

Farmers and landowners – the so-called stakeholders in all of this – asked that the schemes should be broad and shallow. It's certainly shallow, and I'm worried that farmers and landowners who wish to do that bit extra will be restricted in what they can achieve. If we take the proposed arable and horticultural standards then there are three levels of ambition. Under the current pilot scheme, at Intermediate Level – which rewards the farmer with £58 per hectare of grassland – the stated aim is to 'provide resources for birds and pollinators on 8 per cent of eligible land'. These 'resources' for the birds, pollinators and other beneficial insects must include sites for nesting and cover on 2 per cent of eligible land, in addition to habitats rich in insects and flowers on 3 per cent of eligible land. In effect this is saying we will only pay farmers to take 8 per cent of their land out of production and put it into 'public goods'. At Great Farm we have planned to buffer all of our hedgerows with species-rich grassland. That would already take us beyond the 8 per cent of our total acreage, and the excess work and habitat rehabilitation we have done with strips for arable plants and winter food will not receive any further payments.

Another potential weakness in the new proposals is what is known as 'cross compliance'. In the case of hedges, for example, they are protected in England under the 1997 Hedgerow Regulations. A landowner must apply to the local authority if he wishes to remove the hedgerow, which is then supposed to assess the value of the hedge. There is a list of attributes which determine if the hedgerow is important – things such as its age, the number of species within it, and its location. However, since 2003 hedgerows have also been given protection under the notion of 'cross compliance', which aims to ensure that farmers receiving public payments meet a set of environmental standards. They include things like maintaining soil cover (not leaving fields bare), protecting groundwater from pollution and the protection of field boundaries – i.e. not grubbing out hedges and tree lines so as to create giant fields which can be farmed more efficiently.

Hedges, in order to meet agricultural and environmental conditions for boundary features, may not be cut between 1 March and 31 August. Green cover should be maintained within two metres of the centre of the hedge, and cultivation or pesticides and herbicides should not be sprayed within the same distance. However, the RSPB and other environmental groups have highlighted the fact that until the new payment schemes come into force, there is an interim period in which some of these rules may not apply. A significant number of landowners could choose to remove hedgerows to maximise efficiency on their land.

Other elements of the scheme are currently too vague. Farmers seem to be able to interpret the application of higher standards themselves. Take 'soil standards': farmers will be rewarded for improving soil structure, organic matter levels and soil biology. The scheme asks that you increase the organic matter on 20 per cent of your farm. A shrewd farmer would simply find his poorest 20 per cent and put it into a short-term grass ley. This would, technically, improve its organic matter, but it would have limited biodiversity benefits. Why not make provision and indeed encourage farmers to improve the organic matter across the whole of their farm?

There is also the issue of the bureaucracy involved, and how farmers will make the necessary judgements and decisions to make the right environmental improvements. Again, using the example of providing resources for birds and pollinators, the government (Defra) website provides an array of sub-menus alongside its overall guidance. These include how to: create areas of bare ground for invertebrates; provide nesting boxes and wildlife boxes for species at risk and nesting plots for lapwings; allow grass strips; maintain field corners; uselow-input cereal; grow herbal leys; and use unharvested low-input cereal. There are separate sections on creating flower-rich margins, maintaining beetle banks, sowing winter bird food plots in spring, leaving stubbles over winter and creating two-year legume fallows. Another section describes how to create a 'nutrient management plan' for your land. Some of these actions require applying for separate funding under the Countryside Stewardship Capital Grants scheme. The list goes on and on. I can't criticise the ambition in wanting to put this information out there, but finding the time to read all of it and then make decisions about which actions and activities will be most beneficial on one's own land is a mighty task. I suspect that most small farmers will struggle to find the time to do it. So, once again, they will need to rely on agents, agronomists or ecologists to draw up a plan of action. And who will pay for that?

How farmers get rewarded was a key topic within the National Food Strategy plan, where it was recommended that the current agricultural payments system should be extended until at least 2029, to help farmers make the transition to more sustainable land use. Dimbleby's calculations estimated that compensating farmers for improvements to the ecosystem and to supplying the so-called ecosystem services that the Environmental Land Management scheme envisages would cost around £2.5 billion per year, an amount very close to the old EU subsidy. Dimbleby was forthright in his assessment of some of the confusion around the new system, reiterating that in order to meet its net zero commitments (under law), we in the UK would need to change the way our land is used, but that the current strategy is, in his words 'unstructured, unstated and therefore unable to guide good local decision making'. Crucially, it leaves farmers to second-guess the government's priorities, further adding to the uncertainties they have to navigate. Having said all that, the Environment Secretary, George Eustice, has repeatedly said that the gradual phasing out of the old CAP payments system and its replacement with the new environmental schemes will be a process of 'evolution, not revolution', and that the government wants to ensure that in the long term farmers can continue to farm – sustainably. The replacement of the old CAP Basic Payment system will be incremental, and the Sustainable Farming Incentives will gradually swap places with many of the schemes already in operation over a period of several years.

From my viewpoint inside the farming system, it is my firm belief that, as well as requiring a joined-up national overview of how we use our land, any and all of these environmental improvements will only have significant benefits if they are adopted within a whole-farm approach. The current ELM proposals stop short of this. Alongside the whole-farm strategy, we also need to be ensuring that single farms with better environmental approaches are able to coordinate with their neighbours, especially where they may be adjacent to areas of high value, such as national nature reserves, SSSIs, etc. What we don't want to do is set up a system where we have areas that are being farmed highly intensively and areas of high nature value side by side, effectively creating a jigsaw landscape where several pieces are missing. In conservation areas worldwide it has long been established that segregated islands where certain species can survive do not work as effectively as areas joined by wildlife corridors, which allow for the movement of animals and invertebrates which don't have the option of flying from one spot to another.

At the heart of all this, we must expect farmers to be natural capital 'asset managers'. One of the things that I have heard much about recently is that governments simply can't afford the costs that will be involved in paying farmers to restore our countryside. The money needs to come from other sources. The Prince of Wales has even entered this debate, telling the BBC that while governments may bring billions of dollars to the effort, 'the private sector has the power to mobilise trillions of dollars'. His own Terra Carta project (launched in 2021) has the aim of making global investment and financial flows consistent with 'the pathway towards low greenhouse gas emissions, climate-resilient development and Natural Capital/biodiversity restoration'. These aims are connected with his Sustainable Markets Initiative, launched at the World Economic Forum at Davos in 2020. The Prince of Wales is yet another voice within the current debate who believes passionately that 'nature-based solutions' are critical. 'Nature,' he states, 'thanks to the benefit of billions of years of evolution, has already provided us with the solutions. Universal principles rooted in the harmony of nature's patterns, cycles and geometry, which ancient civilisations and indigenous peoples have known all too well, need to be harnessed to inform science, technology, design and engineering, and can drive a sustainable future.' It's a formal way of stating that nature often knows best, and it echoes strongly with what I've seen with my own eyes over a lifetime out and about in the fields, watching how the natural world has the ability to bounce back, to compensate when something in the pattern shifts and changes. The natural world is never static, and even a disrupted seasonal pattern of weather or a flood or a drought will bring unforeseen changes that we barely understand at the time, but which usually act to repair the short-term damage.

Modern farming, of course, relies on money. And farming, whatever its relationship to the natural world, has to be a viable business for those who practise it. If our governments can't afford to pay for the changes they say the planet needs, then it may well be that the business world has to become a more active partner in the natural landscape. Understandably, many farmers are nervous about the prospect of doing deals with the corporate world, and many people, not just farmers, are apprehensive about the idea of corporate entities seeing agriculture as just another investment avenue, or a way of 'greenwashing' their own credentials. If a corporation pays a landowner to, for example, sequester carbon on a piece of land, will that sponsor start demanding a say over what gets planted on it?

I see a lot of potential for corporate partnerships with landowners and farmers, who can enter long-term financial arrangements to restore natural capital for their mutual benefits. But, once again, someone needs to be brokering those deals, and advising farmers about what makes a deal suitable. That expertise is not currently easily available, and there are fears that farmers will have to go through another layer of middlemen to access it. I should say that Defra has brought in a system, albeit small, called the Future Farming Resilience Fund (FFRF) under their Agricultural Transition Plan. The money will go to organisations which support farmers and land managers who are already receiving the Basic Payment Scheme subsidies. Under the FFRF, there is money for these organisations to help an estimated nine thousand farmers and find the ways their businesses can adapt to the new system of payments. This advice will be free to the farmers.

Under the revised terms of the Sustainable Farming Incentive, the government has stated that land can be entered into the agreements even if the land is in a private sector scheme, including such things as carbon trading deals, nutrient trading or biodiversity net gain credits. There is a lot of talk currently about so-called market-based mechanisms and their potential to bring cash into the agricultural system. There are many questions around whether these markets will need to be regulated. Carbon offsetting and sequestration are being posited as areas in which companies, or even industries, can pay to 'offset' their carbon footprints. However, as Professor Dieter Helm has pointed out many times, there is, as yet, no single price for carbon and no transparent market for either carbon offsets or carbon emissions. There are also no common platforms through which offsets or emissions can be traded. Farmers, with all due respect to their talents and skills, are in no place to negotiate this economic minefield at present.

Defra say they will monitor developments in this area, but, in their words, 'will not crowd out private finance'. Defra believes that there is a very low risk of paying out money to farmers for the same sorts of results that the private sector would desire. I'm not the first to say it, but many farmers are distinctly nervous about the longer-term future, and how this involvement with the private sector effectively sponsoring 'ecosystem services' will play out.

13
Small Tweaks

On Great Farm it's still early days, even though we've already seen a big surge in the numbers of breeding pairs of several bird species. I'm confident that nature is gradually starting to fill in the gaps, creating food and nesting habitat, and beginning a ripple of life and energy that will see more and more activity in the growing hedgerows and flower-rich field strips we have created. Variety, heterogeneity, these are the keys to making sure nature has a place on the farm.

Elsewhere on the estate, we are already seeing the results of some of the changes we have made, especially in the management of the grazing marshes. Large, charismatic birds like the spoonbills and the great white egrets have nested, and successfully fledged their chicks. One of the key aspects of what we're doing at Holkham is monitoring our populations. This involves our nature wardens, who have many decades of birdwatching experience, counting individual birds in specific habitats on a regular basis. I am lucky to have Andy Bloomfield and Paul Eele in the roles, both of whom are superlative naturalists, each with their own specialist interests. Andy is an excellent birder, Norfolk born and bred and the son of a Holkham shepherd. He has been especially engaged with monitoring the success of the Holkham spoonbill colony, but he's also an avid spotter of spiders. Andy will happily rise before first light, getting himself into position if there is any chance of spotting something newly arrived. He says he wouldn't want to live anywhere but Norfolk because of the wild geese, migrant songbirds and the new exotics he sees arriving because of climate change. He loves the wildness of the reserve and thinks of it as his back garden. When I asked him what it meant to work here, he said simply that 'the soil of Holkham runs through my veins, and ultimately, when I die, I'll become part of it.' Meanwhile, Paul's ornithological skills are meticulous, and he's very patient and methodical at crunching the numbers. He worked for the RSPB, and his bird-spotting is matched by his knack for discovering rare fungi and knowing where to find emerging orchids. It's thanks to them and their dedication that we have such good data relating to the changes, both positive and negative, in the resident and migratory bird populations on the estate. They are my eyes and ears out on the reserve, and they notice every tiny change in the seasons and remember the precise locations of every natural sign as they move about, largely on foot, each day.

The general improvements to the natural environment around Holkham are being borne out in the individual bird populations. In 2021, the egrets and spoonbills did well across the estate, the latter building forty-five nests, remarkable for such a charismatic bird, which had been extinct as a breeding species in this country since the seventeenth century. The number of nesting pairs at Holkham doubled and, in all, there were almost sixty chicks successfully fledged. That's an impressive increase from the eleven chicks fledged back in 2011, not long after the small colony of birds took up residence. In total, Holkham has seen 454 spoonbill chicks fledged over the last decade.

I think spoonbills tell us a lot about what's happened to our natural environment in this country. The spoonbill is not a phenomenally rare species, and you can find them across Europe, in North Africa and all the way across to Japan. But it's remarkable because it has come back, and without being captured elsewhere and artificially introduced by people. We don't know exactly when the previous colony managed to hatch a chick, but the last recorded English observations were made in 1668, when they made nests at Trimley in the Orwell Estuary, not so very far from Holkham (less than seventy miles as the spoonbill flies). The spoonbill clung on a lot longer than the white storks, which disappeared from England in the fifteenth century. The white storks have been brought back, too, albeit reintroduced as a breeding population, and one of the success stories of the Knepp project has been the hatching of its first chicks in 2020. Along with other small colonies of storks established in Surrey and East Sussex, the project seems to be going well, and when the Knepp storks hatched in a nest in an oak tree, it was the first written record of chicks fledging for 606 years. The biodiversity of the Knepp estate, with its combination of wetlands, grasslands and trees in the open, made for the right habitat. Once again, it's the mixture and the combination of shelter, food and foraging areas which contributes to allowing adults to settle and then successfully rear their young. The same combination of elements is, I'm sure, what has allowed the spoonbills at Holkham to flourish. Both the storks and the spoonbills probably suffered with the progressive draining of East Anglia to turn what were large areas of wetland and marsh into arable land. Both birds were also hunted, of course. These types of landscapes also attract the new and growing populations of egrets – little, cattle and great white – birds which need wetland breeding areas, and food sources.

The spoonbills never abandoned England completely, they would fly in periodically, but just never managed to hatch and rear chicks. The availability of the right type of food for the fledglings is almost certainly a factor. The different types of wetland around Holkham seem to suit them very well, if the burgeoning numbers are any sign. They love a densely vegetated island surrounded by wetlands, and that's exactly what they have at Decoy Wood. It's nice to think that birds which once nested on the Thames at Fulham, and were served up at banquets for Henry VIII, are making a return and are now safe from hunting.

One of the most rewarding sights at Holkham has been the increasing numbers of another, coincidentally, white wading bird, which seems to have particularly thrived on the newly wetted areas inland from the pines. The avocets, like the spoonbills, had been extinct from Norfolk as a breeding population for well over a century, until they returned during the Second World War. It was then that the flooding of the marshes throughout much of East Anglia, which was used as a way of defending the coast, brought the birds back. After a small colony developed on Havergate Island on the River Ore in Suffolk, the island became one of the RSPB's first reserves in 1949. The birds have now re-established themselves in several areas across England, notably along the southern and eastern coastlines. Famously the symbol of the RSPB, these dainty birds have a gently upturned black beak designed to sweep through the shallows in estuaries and sift for small crustaceans, invertebrates and worms.

The avocets are attracted to the freshwater scrape at Wroth's, the large area of muddy shallows south of Decoy Wood also favoured by nesting lapwing. In 2021, across the reserve and the farmland we saw just over a hundred breeding pairs of avocet at Holkham, a significant rise on the eighty-seven pairs observed in the previous year. They no doubt benefited from improved water-level management across the farmland. The 2022 season will see the avocets return to an even more improved freshwater site, thanks to digging carried out at the start of winter. The physical changes we have made to this area are really very small – tweaks rather than major innovations – and go hand in hand with carefully observed monitoring and management of water levels, which can be controlled by the dyke system. It seems highly improbable that for three seasons running we have seen an increase in avocet numbers, but we are getting something right. So far, just three years have brought a 30 per cent increase in avocet breeding pairs.

The rise in lapwings has seen a similar trajectory, with breeding territories within the farmland increasing from 131 in 2020 to 161 in 2021. If we include the coastal strip of the estate then the total number of breeding pairs goes from 208 in 2020 to 256 a year later. Encouragingly, this is a massive increase on the 2019 figure of just 131 territories. This all means that lapwings are now showing numbers that were last seen twenty years ago.

Another species which has shown a significant increase is the redshank, also a wetland breeder on both salt and freshwater marshes. Last year, we had the best year for redshanks in almost a decade, with the bird benefiting from a very wet May. At Overy Marsh, we saw a dramatic rise from just five pairs in 2020, to thirty-seven pairs in 2021. The breeding pairs averaged a breeding success of 0.59 fledged chicks, the best numbers for a decade. Our population of redshank increases dramatically in winter when, like the pink-footed geese, large numbers head south from Iceland for some respite from the arctic conditions. Again, thanks to attentive water-level management, their numbers have risen for three successive winters, and in summer we had 105 breeding pairs.

Sometimes, as I have said, a small tweak in a seemingly simple piece of habitat management can have dramatic effects, whether that is not cutting hedges at a certain time, or making sure that sluice gates are not opened purely by a fixed date in the calendar, but according to weather conditions. Reed buntings, little sparrow-sized birds with distinctive black caps, have seen an increase in breeding pairs, measured by male territories. In 2012 there were a record 102 territories observed, but since then they have declined in most years. One of the contributing factors, in my view, was loss of habitat. As their name suggests, they often nest in reeds, especially the thick growth along the banks of rivers, ditches and dykes. There had been a slightly obsessive regime of clearing out the dykes across Holkham almost annually. We now clear them only when they become obstructed and, sure enough, the last two years have seen record numbers (40 per cent increase) of reed bunting territories established, with a total of 138 singing males seen in 2021.

Not everything is a total win, because in any natural ecosystem change is constant. Coots, moorhens and mallards have all been showing a decline, and many birders around Norfolk are attributing the fall in numbers of these riverine species to the rise in the otter population. Death, predation and the often brutal clearing out of many nests in the space of a few days by an opportunistic predator with its own young to feed are just part of what happens. Sometimes we simply have to stand by and watch.

There is a difference, clearly, when measuring success on the parts of Holkham Estate which, although farmed land, are still part of the National Nature Reserve. There is overlap, of course, but by and large the purely agricultural holdings lie mostly inland, away from the dunes and marshes which attract the charismatic winter visitors in such high numbers: the highly vocal natterjacks and the spectacular breeding colonies of avocet and spoonbills. However, alongside the experiment of returning nature to Great Farm, we have been monitoring the populations of breeding farmland birds on nearby properties. These have included 'in house' (i.e. Holkham-managed) farms as well as tenanted farms. The tenants are free to enter into nature stewardship schemes if they wish to, or not. The holdings surveyed have all been arable farms with short crop rotations, exactly the type of farms where the loss of birds has been greatest in the past fifty years. Just as on Great Farm, baseline surveys on these properties in 2019 revealed low numbers of farmland breeding bird species. For our samples, we recorded blackbirds, chaffinches, linnets, yellowhammers, dunnocks and whitethroats.

One of the in-house farms surveyed was previously signed up to an agri-environment scheme. However, closer inspection revealed that it wasn't delivering very much in the way of species or habitat improvement. While field margins had been left uncut, they had been taken over by brambles and the arable plants had been squeezed out. There were some whitethroats nesting in the bramble, but only one pair each of yellowhammers and linnets, which both need some short grassland to forage for seeds. I recommended cutting back the brambles and cultivating the field margins so that annuals could emerge, while leaving the hedgerows uncut. In 2020, the same field margins saw three breeding pairs of linnets and six pairs of yellowhammers. Skylarks, in the same field, went from six nesting pairs in 2019 to sixteen in 2021. The whitethroats, which had been dominant in the bramble, saw a decline from twelve pairs in 2019 to just four pairs in 2020. However, as the brambles recovered, but were only allowed to grow back in controlled areas, the whitethroat population had recovered to eleven pairs by 2021. Across this same farm, the number of breeding chaffinch pairs doubled over the three summers. Similarly, breeding pairs of dunnocks (a dainty little bird sometimes called a 'hedge sparrow') also tripled within one year, and maintained the same numbers the following year. The number of individual birds may seem small, but imagine doubling or tripling the number of these individual species on all – or even half – of the farms in England. With the continued provision of the most suitable breeding areas we would, for the first time in three farming generations, allow these birds to bounce back.

When we surveyed one of the tenanted farms, it showed a slightly healthier baseline, however it was not part of any agri-environment scheme. After friendly conversations with the farmer, he signed up to one of the Defra programmes. I proffered some advice on cultivating his field margins. Having had no breeding whitethroats recorded on the farm, in 2021 they saw the arrival of four nesting pairs. Skylarks, which had been minimally present (two breeding pairs in 2019), had risen to fifteen pairs in 2021.

I am heartened when a farmer rings me to say they have just spotted a bird or, sometimes, a plant on their land which they have not seen for years. If I sometimes seem critical of farmers' attitudes to the natural environment, it is only because I have seen, all too often, how many of them have been pushed into regarding their land as purely a business, from which profit must be extracted at almost any cost. Under the old system, they worked hard for pretty meagre returns. Almost my entire working life has been spent with farmers, and I truly believe I understand them. I know how hard it is to work the land and be at the mercy of the seasons. I see the frustrations they experience in being given new directives by the government, or by wholesale food buyers who have very narrow parameters into which the farmer must fit – including diktats about the size and shape of root vegetables. But I also know how inventive and patient farmers are. I believe that those who take the lead on the new environmental style of farming will do well. If they use the skills and the willingness to innovate that they employ in the difficult business of producing food, then I believe they will also become world leaders in delivering a healthy environment and natural world for everyone.

14
Pink Feet

Holkham
December

The light changes fast as dawn approaches, when an icy-blue paling of the sky is a precursor to the arrival of the wintry sun. As if a dimmer switch is being slowly turned, the shapes in the land start to emerge. The pines rise from the dunes and the faint outlines of the cattle begin to move gently as they too stir for the day. It's peaceful, a time filled with the natural rhythm of the earth, until the stillness is interrupted by a hoarse cry somewhere on the reserve. It's an unearthly half-bark, half-scream, unsettling, with a stabbing angry quality that is eerily human. Muntjac.

As the light strengthens, there is another sound: a succession of high-pitched cackles as somewhere out on the ponds at Wroth's the geese begin to talk to one another. Two birds rise from the shadowy ground and fly towards me, greylags, with a deeper, guttural call. On the ground the chatter is rising, but as yet nothing has taken to the air. There will come a moment, unpredictable but inevitable, when the talkative visitors from the far north Atlantic will set off for the day's foraging. The morning chill will not wear off today, a cloudless sky offering no insulation and the low sun giving very little warmth. Minutes pass. The hubbub of voices builds but then, suddenly, there is a moment of silence. It's then that the first line of birds takes off, numbering just forty or fifty at first. And then comes another wave. The noise is like a dozen squeaky wheelbarrows trundling along, and then a hundred more. Some unseen, mysterious command has gone out and the geese are taking to the air all at once. A dark cloud rises from the marsh. Too many to count. The squeaking, shrill calls are blending like a choir, blotting out the wind and any other noise. There are thousands of them. And from the east, nearer to Wells, there are more, higher and already strung out into skeins, the distinctive V-shaped formations that their dark silhouettes draw against the early morning sky. They are purposefully flying south and west, in search of food. From afar, the crowd of birds is so dense that it looks like a murmuration of starlings, but these geese are bigger than mallards, and not much smaller than a swan. They have journeyed 1,500 miles or more to escape the cold Icelandic winter, where there will be very little for them to eat for several months. Even so, a small number – around 5 per cent of the pink-foot population – do stay in Iceland, wintering around the coastline rather than in the Highland breeding areas.

The rising flights of pink-feet are one of the great

natural spectacles of British nature. In the past fifty years, the numbers of these geese visiting Norfolk has been steadily increasing, and in this area we see upwards of 100,000 now. They also overwinter in Scotland, where massive gatherings of over 50,000 individual birds have occasionally been seen. In 2016, a record individual flock of 90,000 individuals arrived at the Montrose Basin, north of Dundee, a regular staging post for the geese as they come south. They'll stay with us until March, and some perhaps even up to May, if we're lucky.

In the Arctic, these birds have been extensively studied and monitored, and we know there are two distinct populations – those from Iceland and Greenland and another, smaller group, which summers in Spitsbergen in the vast Svalbard archipelago at more than 78° north. There, the pink-feet are the commonest species of goose, with more than 80,000 geese mainly around the fjords of Spitsbergen. They spend the winter on the continent in Belgium, Netherlands and Denmark. The Icelandic population is much larger and is regularly measured as part of the Icelandic-Breeding Goose Census, an international effort which now puts their population at around half a million birds. This is a significant increase since the 1960s when there were only about 60,000 birds. Around 25,000 pink-feet are shot by hunters each year in Iceland.

The pink-feet that reach Norfolk have bred during the summer in the Icelandic Highlands, massing in huge flocks where they cooperate to fend off predators and feed on the tundra plants like sedges and mosses, sorrels and crowberries. When they arrive back in Iceland or Svalbard in spring, there is a good chance that they will find their feeding grounds still covered in snow. Studies have shown that they will cope, favouring marshy areas where there is water available to drink as well as freshly shooting 'horse hair' (*Equisetum*), a prehistoric rush which is one of the first species to emerge after the winter freeze. This is also the type of habitat where the ground will be softened by glacier melt, and where they can successfully grub for underground plant stems and rhizomes. The rise in populations over the last decade seems likely to be linked to increased food availability in the far north, which in turn is highly likely to be linked to climate change and warmer cycles with less snow and ice, as is now commonly observed. Their favoured breeding grounds have traditionally been the high marshes of Hofjökul and Eyjabakkar in the Vatnajökull National Park, at the fringes of Iceland's largest glacier, where the land is filled with mud flats and ponds. Again, due to changes in climate the geese are now increasingly colonising lowland areas too.

Just before winter sets in, the pink-feet moult and lose the ability to fly, and seek safety by spending almost a month close to coastal waters where they can escape from terrestrial predators. Arctic foxes are the main threat on land, but chicks can be taken by skuas, gulls and falcons. When the birds can fly again, and as the first snows threaten, they head south, with a few making it to Ireland, but the majority touching down in Scotland and at sites like the Ribble Estuary in the north of England, before the numbers here in Norfolk gradually build. They come here in search of grass and supplement their diet with high-carbohydrate foods like sugar beet tops, potatoes and oil seed rape. The pink-feet's talent for grubbing the soil and uncovering root crops comes to the fore.

Ten hours after the mass flight of the geese at dawn, I am driving along the road I call 'Taliban Alley', west of Wighton. It's still Holkham land, but it's part of an area where the farms have been in the hands of older tenants, the generation on the point of retiring, who have been immersed in the high-productivity, high-input attitudes that I have previously described. There is a slight rise in the land, and the hedges both beside the road and dividing the crop fields have been shorn and slashed into neat rectangular lines. Many of these hedge species – which stubbornly, miraculously, somehow cling to life – are little more than ivy-covered stumps, physical demarcations in the land but of almost no biodiversity value, a useless habitat for feeding or nesting birds. In fairness to the farmers, many of these hedges were planted during the Inclosure Acts period and they contain just one or two species of plant. Such hedges were usually created by planting hawthorn and/or blackthorn. One formula, devised in the 1970s by Dr Max Hooper, says that you can age a hedge by counting the species of shrub in a thirty-metre section – one extra new species will be found for every century it has been growing. Hooper surveyed hundreds of hedges with recorded ages of between 75 and 1,100 years. This is a rule of thumb, rather than an infallible method because, if left alone, the right habitat will allow several species to colonise ground and create a hedge in a much shorter space of time During the Inclosures period, most hedges were planted with just one species; older hedges, however, often had more species present right from the beginning. Also, the older a hedge is, the more likely it is to have had more species in it, and the more likely it is to be a natural landscape feature. In the north of England, Hooper's rule is less applicable, due to a lower range of species which thrive in the cooler climate. However, the fact that this age calculation method arose, and is applicable in much of England, is due to how intensively managed the British countryside has been over the past several centuries.

The land I can see on either side of the road is bare, with no winter cover crop emerging. It's typical of land that has been in an intensive rotation, relentlessly planted with cereals used for cattle feed. It's a monoculture on some of these farms, leaving no space for nature and making the land a desert for wildlife in the winter. All too often, farming this way is a symptom of our obsession with eating meat – or at least cheap meat – and too much of it. On my right is a harvested potato field, which looks like part of the Somme, ploughed up and now waterlogged. Further up the road is a field of beets where about a hundred pink-feet are foraging. It's an exposed place and there is a steady drizzle. It's a dismal sight, these elegant birds trudging through the mud, grubbing through the wet soil for the discarded tops of the roots. Pink-feet are nervous of people and approaching them isn't easy. Here, with no vegetation obscuring my view, this is a good chance to see them well. Their dark head feathers are a fine shade of russet, melding with the striated feathers of the neck. A pale breast and white rump sit atop those distinctive pale pink feet and legs, and their stubby beak is also pink with a dark base and tip. The greylags, which often inhabit the same territory and also visit from Iceland, have much paler colouring and a yellow beak, and they are larger than the pink-feet overall.

I head home, making a detour back via the coastal road where I can watch the sunset. The rain has passed and the sky is clear again, save for a dark line of thinning clouds here and there. The official time for sunset passes too, but there is still a faint wash in the sky. Far off to the west, the line of clouds turns pink before the horizon shimmers that cold neon blue for a final few minutes, before the real darkness sets in. The pink-footed geese have timed it perfectly, and there is just enough distinction between land and sky for me to spot the cloud of birds heading back to the roost. They come in waves, as they did at dawn, but they are tightly packed and descend on the scrape between me and the pines in less than a minute. The screeching kerfuffle as they land is just as loud as it was at dawn, like a slowed-down version of a braying donkey. Wheeling and landing into the slight breeze, they somehow find their spot for the night. Within moments, there is almost no sound, and it is as if they were never there.

Acknowledgements

Jini Fiennes – for the endless encouragement
Mark Fiennes – for the detail and perfection
Teale and Nathaniel Fiennes – for a reason
Barbara Linsley – for the unconditional support
John Horsman – for the example
Charlie Burrell – for the break
Isabella Tree – for the passion about nature
Nicholas Bacon – for the faith in me
Tom Leicester – for the opportunity
Tim Ecott – for the words
Sam Knight – for the 'fucks'
The farmers, foresters, gamekeepers, gardeners, wardens and rangers – for the knowledge and expertise they shared